Quantitative Genetics and Crop Breeding

nipa
Browse Subject
Agriculture Sciences (679)
Animal Husbandry and Veterinary Sciences (261)
Climate Change and Environmental Sciences (63)
Community Science (Home Science) (39)
Competitive Examinations and MCQ's (57)
Dairy Technology (91)
Fisheries Sciences (26)
Food Science and Technology (155)
Geological Sciences (50)
Horticultural Sciences (342)
Plant Sciences (191)
Skill and Entrepreneurship Development (47)
eBooks / eChapters / Articles
eBooks
Explore Now
Browse, Search, Read & Buy...
Subject Catalogues
eChapters
Publishers
Print Books
Forthcoming
eBooks
New Titles
scan for catalogue
NIPA
GENX
ONLINE RESOURCES
Publishing Books and Journals
Ebooks and e articles, Current Affairs
Reasoning, Logic and Aptitude
Competitive Examination Preparation
Language Learning and Development Programme
Document Quality Checker and Improvement Tool
Effective Public Speaking, Presentation and Interpersonal Skills
Online Programmes for Professional Development
Personality Development and Human Values
PAY USING
UPI
PayPal

Quantitative Genetics and Crop Breeding

By :

S. THIRUGNANAKUMAR

K. SARAVANAN

N. SENTHILKUMAR

A. ANANDAN

R. ESWARAN

2012

New India Publishing Agency

Pitam Pura, New Delhi-110 088

Published by
Sumit Pal Jain *for*
New India Publishing Agency
101, Vikas Surya Plaza, CU Block, L.S.C. Mkt.,
Pitam Pura, New Delhi- 110 088, (India)
Phone : 011-27341717, Fax: 011-27341616
Mobile : 09717133558
E-mail : info@nipabooks.com
Web : www.nipabooks.com

ISBN : 978-93-80235-98-1

Typeset at: Harminder Singh Kharb *for* Laxmi Art Creation

Printed at: Jai Bharat Printing Press, Delhi

PREFACE

The father of genetics, Gregor Johann Mendel proposed the laws of inheritance based on his experiments in pea crop. The important laws are: 1) Law of segregation or purity of gametes and 2) Law of independent assortment. Mendel had two important assumptions *viz.*, 1) a heterozygote produces two types of gametes in equal proportion and 2) gametes fuse together randomly. The two laws are valid till date. Interestingly, the segregation ratio obtained by Mendel in his experiments are similar to the ratios obtained using probability principles. Thus, Mendel recognized the statistical nature of the genetical variables (Singh and Chaudhary, 1977). Later, Fisher, Sewall Wright and Haldane developed the generalized theories and a separate discipline namely Quantitative Genetics bloomed.

In Quantitative Genetics, the measurable characters are considered. Continuous variation is the characteristic feature of a quantitative trait. Quantitative genetics perceive genes based on phenotypic expression. Johannsen (1909) proposed the concept of genotype and phenotype. Multiple factor hypothesis was proposed by Nilson-Ehle (1909); Emerson and East (1913) and Davenport (1913). The work of Mather and his co-workers, Comstock, Robinson Cockerham, Kempthorne, Hayman, Jinks etc. in plants and Lush and his co-workers in animals gave further impetus to the study of quantitative genetics. Quantitative Genetics provides the foundation for breeding methodologies.

Several books are available in Quantitative or biometrical genetics. However, the present work is unique in that sense it gives formulae along with actual data (analyzed) for the easy understanding. This book is mainly meant for post graduate and research scholars in Quantitative

Genetics. A careful perusal of the book will give clear cut idea about the interpretation of the data and formulation of breeding strategies. However, the authors ready to accept ideas and suggestions to improve the quality of the book. The ultimate aim of the authors is to make the scholars to write dissertation (thesis), project reports and articles without any hurdle. The authors wish all the best for the readers.

Authors

CONTENTS

Preface v

Chapter 1
Genetic Variation 1
1.1. Analysis of variance (ANOVA), 2
1.2. Example, 3

Chapter 2
Correlation and Causation 11
2.1. Correlation studies, 12
2.2. Path analysis, 20

Chapter 3
Discriminant Function Analysis 33
3.1. The analysis, 34
3.2. Example, 35

Chapter 4
Stability Analysis 41
4.1. Stability parameters, 43
4.2. Analysis of variance for phenotypic stability, 44
4.3. Test of significance, 45
4.4. Simple correlation coefficients, 45
4.5. Example, 45

Chapter 5

Genetic Divergence 57

5.1. Wilk's lambda criterion, 58

5.2. D^2 analysis, 59

5.3. Intra and inter-cluster distances, 59

5.4. Example, 60

Chapter 6

Combining Ability 77

6.1. Line x tester analysis, 77

6.2. Diallel analysis, 91

6.3. Triallel analysis, 111

6.4. Quadriallel analysis, 128

6.5. Heterosis, 134

Chapter 7

Genetic Analysis of F_1 Generation 145

Chapter 8

Graphic Analysis in F_1 Generation 151

Chapter 9

Genetic Analysis of Early Segregating Generations 163

9.1. Generation mean analysis, 163

9.2. Statistical analysis, 165

9.3. Example, 168

Chapter 10

Genetic Analysis of Back Cross F_1 Populations 179

10.1. Generation mean analysis, 179

10.2. Estimation of genetic effects m, [d] and [h] on the assumption of the simple additive-dominance model, 180

10.3. Estimation of genetic effects m, [d], [h], [i], [j] and [l] on the assumption of the presence of digenic interaction, 181

10.4. Partitioning the variance components, 182

10.5. Example, 183

10.6. Summary of the results on gene effects, 183

Chapter 11
Genetic Analysis of Triple Test Cross Progenies.... 199
11.1. Statistical analysis, 199
11.2. Example, 201

Chapter 12
Genetic Analysis of Biparental Progenies 211
12.1. Statistical analysis, 213
12.2. Example, 214

References 223

Index 239

Chapter 1

Genetic Variation

Any crop improvement programme primarily depends on the amount of genetic variability available and the extent to which the economic traits are heritable. A wide survey of genetic variability and a thorough understanding of the genetic makeup of the crop with the biometrial tools is very much indispensable for initiating an effective breeding programme.

Effectiveness in breeding programme depends upon three main factors *viz.,* (i) the extent of genetic variation available, (ii) heritabiity of the variation and (iii) intensity of the selection pressure that can be applied. As most of the characters of economic importance and related characters which influence the performance of the economic character in a given set of environment are highly influenced by the environmental conditions, the heritable (genotypic) variation usually masked by non-heritable (environmental) variation and thereby creates difficulty in exercising the selection. Hence, it becomes necessary to partition the overall variability into heritable and non-heritable components to enable the breeders to plan for proper breeding programme.

The estimates of mean, variance and standard error are worked out by adopting the standard methods of Panse and Sukhatme (1961). Usually, randomized block design is the choice of the statistical design.

Various genetic parameters are worked out from the mean squares. Assessment of genetic variation is usually made through the computation of genotypic variance and genotypic coefficient of variation. The proportion of the total genetic variance expressed as percentage of total phenotypic variance has been used as a measure of genetic variability of heritability in broad sense.

1.1. Analysis of variance (ANOVA)

The data from replicated trials is analyzed in randomized block design (RBD) and analysis of variance (ANOVA) is constructed. The ANOVA Table 1 illustrates clearly, whether the genotypes taken for the study is genetically different with reference to the characters studied. The expectation of ANOVA table of RBD is given below.

Table 1 : ANOVA (single environment or season or location)

S. No.	Source	Degrees of freedom (df)	Mean squares (MS)	Expectation of mean squares
1.	Replication	(r – 1)		
2.	Genotypes	(g-1)	M_1	$\sigma^2e + r\sigma^2g$
3.	Error	(r-1) (g-1)	M_2	σ^2e
4.	Total	(rt – 1)		

(Source: Panse and Sukhatme, 1961)

Where,

r = number of replications

g = number of genotypes

M_1 = mean squares for genotypes

M_2 = mean squares for error

σ^2g = genotypic variance

σ^2e = error variance

The significance test was carried out by referring to the standard 'F' table at genotypic and error degrees of freedom, given by Snedecor and Cochran (1961).

The ANOVA for the experiments conducted over more than one environment or season or location is given below (Panse and Sukhatme, 1961) (Table 2).

Table 2 : ANOVA (more than one environment or season or location)

S. No.	Source	Df	MS	Expected MS
1.	Block with in season	l(b-1)	-	-
2.	Season (l)	(l-1)	-	-
3.	Genotype (g)	(g-1)	M_g	$\sigma^2 e + b\sigma^2_{gl} + bl\sigma^2_g$
4.	Genotype × season	(g-1) × (l-1)	M_{gl}	$\sigma^2 e + b\sigma^2_{gl}$
5.	Pooled error	n (b-1) × (g-1)	M_e	$\sigma^2 e$

(Source: Panse and Sukhatme, 1961)

The σ^2_g (genotypic component) and σ^2_{gl} (genotype × season interaction) estimates were obtained as follows:

$$\sigma^2_g = \frac{M_g - M_{gl}}{bl}$$

$$\sigma^2_{gl} = \frac{M_{gl} - Me}{b}$$

where, b = number of blocks, l = number of seasons

1.2. Example

The data obtained by Thirugnanakumar (1991) on studying 60 genotypes of sesame (three replications) over three different seasons (SI, SII, SIII) for 16 seed yield and their contributing characters are furnished in Table 3. The characters observed are: X_1 = Days to maturity, X_2 = Plant height at maturity, X_3 = Number of branches, X_4 = Number of flowers, X_5 = Number of capsules, X_6 = Volume of capsules (cm^3), X_7 = Number of seeds per capsule, X_8 = Length of seeds (mm), X_9 = Breadth of seeds (mm), X_{10} = 1000 seed weight (g), X_{11} = 1000 seed volume (cc), X_{12} = Seed density (cc/g), X_{13} = Oil content (%), X_{14} = Seed yield per plant (g), X_{15} = Total dry matter production (g), X_{16} = Harvest index (%).

The analysis of variance (Table 3), for individual seasons and pooled over different seasons indicated significant differences among the 60 genotypes for all the 16 characters studied, except for plant height at maturity in season I (SI) and pooled analysis (P). This explained that the genotypes selected for the study are distinctly different and hence, further analysis is appropriate.

Table 3 : Analysis of variance for individual seasons and pooled analysis over seasons for different characters

S. No.	Characters	Mean squares								"F" values			
		Treatments				Error							
		SI (df .59)	SII (df .59)	SIII (df .59)	ρ (df .59)	SI (df .59)	SII (df .59)	SIII (df .59)	ρ (df .177)	SI	SII	SIII	ρ
1.	X_1	15.5360	11.2987	82.6441	55.27	0.5148	0.3008	1.0826	0.640	30.177*	37.5563**	76.3366**	86.77**
2.	X_2	6158.1300	174.2145	261.4613	2511.69	5442.9840	8.2225	33.7638	1828.330	1.1314	21.1876**	7.7438**	1.37
3.	X_3	16.2457	4.5825	6.1539	9.46	0.6535	0.5020	1.0759	0.740	24.8601**	9.1278**	5.7199**	12.71**
4.	X_4	1723.2550	486.1716	804.0074	1526.91	51.4428	11.4799	14.1112	25.670	33.4985**	42.3499**	56.9764**	59.48**
5.	X_5	1472.6720	537.6636	1426.5880	1743.29	54.4375	7.2426	16.9068	26.200	27.0525**	74.2364**	84.3796**	66.54**
6.	X_6	0.0148	0.0188	0.0135	0.02	0.0006	0.0006	0.0006	0.001	23.7164**	29.5072**	20.8792**	33.04**
7.	X_7	17.6202	6.4264	3.4439	9.19	5.0434	0.625	1.2648	2.310	3.4937**	10.2822**	2.7228**	3.98**
8.	X_8	0.0591	0.0694	0.0690	0.09	0.001	0.0018	0.0065	0.003	60.2215**	38.1661**	10.6201**	28.81**
9.	X_9	0.0151	0.0250	0.0270	0.02	0.0006	0.0018	0.0075	0.003	26.8468**	14.1463**	3.5822**	6.05**
10.	X_{10}	0.3328	0.2314	0.9726	0.47	0.0078	0.0120	0.0635	0.030	42.8710**	19.2401**	15.3092**	16.75**
11.	X_{11}	9.3194	0.2831	9.9527	13.18	0.0338	0.496	0.0515	0.050	276.0679**	5.6856**	193.3136**	292.76**
12.	X_{12}	0.0433	0.0123	0.0584	0.06	0.0004	0.0019	0.0021	0.002	103.642**	6.2985**	27.4749**	39.60**
13.	X_{13}	14.1936	16.1862	21.3146	25.35	0.3692	0.2778	0.2423	0.300	38.4469**	58.2641**	87.9607**	85.50**
14.	X_{14}	6.6522	6.0107	23.1833	17.68	0.7751	1.1822	0.3692	0.780	8.5819**	5.0843**	62.7936**	22.79**
15.	X_{15}	70.9347	64.1904	120.5031	121.40	11.6571	11.5454	4.1092	9.100	6.0851**	5.5598**	29.3254**	13.33**
16.	X_{16}	97.3091	143.5399	148.8468	186.50	7.3211	9.6642	7.0456	8.010	13.2916**	14.8528**	21.1264**	23.28**

* Significant at P = 5% ** Significant at P = 1% (Source: Thirugnanakumar, 1991)

1.2.1. Mean expression

The mean data (over three replications) on seed yield of 60 genotypes (G1-G60) studied over three different seasons are tabulated (Table 4).

Table 4 : Mean expression of 60 genotypes of sesame for seed yield per plant

G. No.	S I	S II	S III	Pooled	G. No.	S I	S II	S III	Pooled
G1	2.94	5.32	3.46	3.90	G31	6.82**	5.36	17.58**	9.92**
G2	4.26	5.88	4.56	4.90	G32	4.01	6.99	11.14**	7.38**
G3	7.87**	4.73	4.18	5.59	G33	4.64	4.00	5.93	4.86
G4	5.22	4.15	4.44	4.60	G34	5.05	6.61	10.48**	7.38**
G5	4.45	4.74	5.20	4.79	G35	4.73	4.98	7.41*	5.71
G6	6.05	3.32	3.50	4.29	G36	5.34	3.81	6.57	5.24
G7	4.98	8.00**	7.98**	6.99**	G37	2.86	4.70	6.96	4.84
G8	4.14	3.90	3.67	3.91	G38	4.36	5.76	8.45**	6.19
G9	3.86	5.37	5.01	4.75	G39	4.77	6.89	7.66**	6.44
G10	4.43	6.00	3.46	4.63	G40	10.07**	5.49	1.62	5.73
G11	9.31**	7.36*	5.74	7.47**	G41	1.58	1.43	2.28	1.76
G12	5.97	6.74	10.88**	7.86**	G42	3.11	2.13	3.04	2.76
G13	3.04	6.55	2.10	3.89	G43	1.63	3.14	3.21	2.66
G14	4.91	6.33	0.68	3.97	G44	4.06	7.29*	18.31**	9.88**
G15	6.16	4.73	1.66	4.18	G45	9.93**	5.51	1.79	5.74
G16	5.54	9.89**	7.77**	7.73**	G46	2.63	2.42	2.68	2.58
G17	3.96	3.07	0.62	2.55	G47	1.40	4.60	5.98	3.99
G18	3.09	6.75	4.59	4.81	G48	2.49	3.55	3.19	3.08
G19	2.76	4.58	3.30	3.55	G49	5.60	3.70	4.27	4.52
G20	3.59	4.91	13.82**	7.44**	G50	2.56	7.29*	6.94	5.59
G21	2.42	2.99	3.12	2.84	G51	2.94	3.65	3.41	3.33
G22	3.53	4.59	6.98	5.03	G52	4.15	5.73	4.14	4.67
G23	4.21	3.31	3.78	3.77	G53	4.69	2.45	4.66	3.93
G24	4.03	5.87	5.58	5.16	G54	3.24	3.16	0.99	2.46
G25	1.78	2.73	4.03	2.85	G55	4.06	2.53	3.56	3.38
G26	5.89	6.80	12.20**	8.29**	G56	2.85	4.63	3.26	3.58
G27	4.99	6.80	10.09**	7.29**	G57	2.46	4.33	6.43	4.41
G28	3.37	7.42*	5.77	5.52	G58	4.49	6.70	6.14	5.78
G29	3.56	5.62	4.44	4.54	G59	4.79	3.83	3.92	4.18
G30	5.98	6.41	10.20**	7.53**	G60	5.02	2.28	11.49**	6.26
					Mean	**4.38**	**4.99**	**5.77**	**5.05**

(Source: Thirugnanakumar, 1991).

	S I	S II	S III	p	
SE	0.62	0.77	0.43	0.51	at p = 5%
CD	1.76	2.18	1.22	1.42	at p = 1%

None of the genotypes consistently showed higher mean seed yield (higher than grand mean + 1CD). Hence, the genotypes which consistently performed with higher mean seed yield in any two out of the three seasons evaluated could be suggested for further breeding programme. Accordingly, the genotypes *viz.*, G7, (SII and SIII); G11 (SI and SII); G16 (SII and SIII), G31 (SI and SIII) and G44 (SII and SIII) are suggested. The measures in different seasons may cancel each other in pooled analysis and hence it is not advisable to take into the consideration of the mean performance.

1.2.2. Phenotypic and genotypic variances

These were estimated according to the method of Lush (1940).

$$\text{Genotypic variance } (\sigma^2_g) = \frac{M_1 - M_2}{r}$$

$$\text{Phenotypic variance } (\sigma^2_P) = \sigma^2_g + \sigma^2_e$$

The Table 5 furnishes the phenotypic and genotypic variance. The phenotypic variance were higher than the genotypic variance for almost all the 16 characters. However, the major part of the variances was contributed by the genotypic component and the environmental component was less. The genotypic variance was higher for plant height (X_2), number of flowers (x_4), number of capsules (X_5), total dry matter production (X_{15}) and harvest index (X_{16}).

1.2.3. Phenotypic and genotypic coefficient of variation (PCV and GCV)

These are computed according to Burton (1952), as follows:

$$PCV = \frac{(\sigma^2_p)^{1/2}}{\overline{X}} \times 100$$

$$GCV = \frac{(\sigma^2_g)^{1/2}}{\overline{X}} \times 100$$

where,

$(\sigma^2_P)^{1/2}$ = Phenotypic standard deviation

$(\sigma^2_g)^{1/2}$ = Genotypic standard deviation

$\overline{X}$ = Grand mean

Table 5 : Genotypic and phenotypic variance for 16 characters in sesame studied over three different seasons (SI, SII, SIII) and pooled analysis (P)

S. No.	Characters	Variance							
		Phenotypic				Genotypic			
		S I	S II	S III	P	S I	S II	S III	P
1.	X_1	8.03	5.80	41.86	9.74	7.51	5.50	40.78	9.11
2.	X_2	5800.50	91.22	147.62	1942.22	357.59	82.99	113.85	113.89
3.	X_3	8.45	2.54	3.61	2.20	7.80	2.04	2.54	1.45
4.	X_4	887.32	248.82	409.05	275.88	835.90	237.34	394.94	250.21
5.	X_5	763.58	272.45	721.73	312.38	709.10	265.20	704.85	286.18
6.	X_6	0.0077	0.0097	0.0071	0.0040	0.0071	0.0091	0.0064	0.0030
7.	X_7	11.33	3.53	2.35	3.45	6.29	2.90	1.09	1.15
8.	X_8	0.0299	0.0356	0.0376	0.0174	0.0289	0.0338	0.313	0.0143
9.	X_9	0.0078	0.0134	0.0172	0.0061	0.0073	0.0116	0.0097	0.0028
10.	X_{10}	0.17	0.1217	0.518	0.10	0.16	0.1097	0.454	0.07
11.	X_{11}	4.68	0.1665	5.004	2.23	4.64	0.1166	4.95	2.19
12.	X_{12}	0.0219	0.0071	0.0303	0.0111	0.0215	0.052	0.0282	0.0096
13.	X_{13}	7.28	8.23	10.78	4.47	6.91	7.95	10.54	4.18
14.	X_{14}	3.71	3.59	11.78	3.59	2.94	2.41	11.40	2.82
15.	X_{15}	41.29	37.87	62.30	27.82	29.64	26.33	58.20	18.72
16.	X_{16}	52.32	76.60	77.95	37.76	45.00	66.95	70.90	29.75

(Source: Thirugnanakumar, 1991).

The Table 6 exhibits the PCV and GCV values for 16 traits in sesame, derived over three different seasons and pooled analysis. PCV and GCV values are unit free and hence comparable. In general, all the traits had higher PCV and GCV. However, there is close agreement between PCV and GCV, illustrating the less influence of environment over the expression of the 16 traits given in the Table 6. The GCV was higher for X_4 (number of flowers), X_5 (number of capsules), seed yield (X_{14}), total dry matter production (X_{15}) and harvest index (X_{16}). The high GCV estimates for the characters listed in Table 6, indicates that there is scope for improvement by selection.

Table 6 : PCV and GCV for 16 traits in sesame studied over three different seasons and pooled analysis

S. No.	Characters	Variance							
		Phenotypic				Genotypic			
		S I	S II	S III	ρ	S I	S II	S III	ρ
1.	X_1	8.03	5.80	41.86	9.74	7.51	5.50	40.78	9.11
2.	X_2	5800.50	91.22	147.62	1942.22	357.59	82.99	113.85	113.89
3.	X_3	8.45	2.54	3.61	2.20	7.80	2.04	2.54	1.45
4.	X_4	887.32	248.82	409.05	275.88	835.90	237.34	394.94	250.21
5.	X_5	763.58	272.45	721.73	312.38	709.10	265.20	704.85	286.18
6.	X_6	0.0077	0.0097	0.0071	0.0040	0.0071	0.0091	0.0064	0.0030
7.	X_7	11.33	3.53	2.35	3.46	6.29	2.90	1.09	1.15
8.	X_8	0.0299	0.0356	0.0376	0.0174	0.0289	0.0338	0.0313	0.0143
9.	X_9	0.0078	0.0134	0.0172	0.0061	0.0073	0.0116	0.0097	0.0028
10.	X_{10}	0.17	0.1217	0.518	0.10	0.16	0.1097	0.454	0.07
11.	X_{11}	4.68	0.1665	5.004	2.23	4.64	0.1166	4.95	2.19
12.	X_{12}	0.0219	0.0071	0.0303	0.0111	0.0215	0.052	0.0282	0.0096
13.	X_{13}	7.28	8.23	10.78	4.47	6.91	7.95	10.54	4.18
14.	X_{14}	3.71	3.59	11.78	3.59	2.94	2.41	11.40	2.82
15.	X_{15}	41.29	37.87	62.30	27.82	29.64	26.33	58.20	18.72
16.	X_{16}	52.32	76.60	77.95	37.76	45.00	66.95	70.90	29.75

(Source: Thirugnanakumar, 1991).

1.2.6. Heritability and genetic advance

Heritability is the extent of transmission of the character from generation to generation. The magnitude of heritability is influenced by the variability existing between the populations and the extent to which a particular character is influenced by the following environmental conditions. Thus, the observed variability for different characters in a crop is partly heritable (due to genetic constitution) and partly non-heritable (due to environment). So, broad sense heritability estimates are calculated in order to estimate the proportion of the total variance attributed to the genotypic differences.

Heritability in the broad sense is calculated according to Lush (1940) and expressed in percentage.

$$h^2 \text{ (broad sense) } \% = \frac{\sigma_g^2}{\sigma_p^2} \times 100$$

The heritability estimates that is above 30-60% are considered as high values, according to Robinson (1966).

Genetic advance is calculated to predict the net effect of selection, as the heritability estimates alone do not provide sufficient information for the genetic improvement that would result from the selection of best individuals. Ramanujam and Tirumalachari (1967) opined that broad sense heritability includes both additive and epistasis gene effects and hence such heritability estimates would be reliable, if they go on hand with genetic advance.

Genetic advance is worked out based on the formula given by Johnston *et al.* (1955).

$$\text{Genetic advance (GA)} = h^2 \times \sigma p \times k$$

where

k = selection differential at 5% selection intensity, which is equal to 2.06

Genetic advance is usually expressed as per cent of mean.

$$\text{GA as \% of mean} = \frac{\text{GA}}{\text{Grand mean}} \times 100$$

The heritability estimates and genetic advance is presented in Table 7. High heritability estimates were consistently observed for all the characters listed in Table 7, except for plant height (X_2), which recorded low estimates in SI and pooled analysis. The genetic advance as per cent of mean was high (above 30 per cent) for seed yield (X_{14}), number of capsules (X_5), 1000 seed volume (X_{11}), harvest index (X_{16}), number of flowers (X_4), total dry matter production (X_{15}), seed density (X_{12}) and capsule volume (X_6). High heritability estimates coupled with high genetic advance as per cent of mean was observed for number of flowers (X_4), number of capsules (X_5), capsule volume (X_6), volume of 1000 seeds (X_{11}), seed density (X_{12}), seed yield (X_{14}), total dry matter production (X_{15}) and harvest index (X_{16}). These characters may be controlled by additive gene action (Johnson *et al.*, 1955). Hence, resorting to simple selection would be rewarding.

Table 7 : Heritability estimates and genetic advance for 16 traits in sesame

S. No.	Characters	Heritability (Per cent)				Genetic advance				Genetic advance as per cent of mean			
		S I	S II	S III	ρ	S I	S II	S III	ρ	S I	S II	S III	ρ
1.	X_1	93.59	94.81	97.41	93.50	5.47	4.71	13.00	6.02	5.14	4.50	13.32	5.85
2.	X_2	6.16	90.99	77.13	5.86	9.68	17.92	19.32	5.33	11.31	34.22	38.59	8.50
3.	X_3	92.77	80.25	70.24	66.13	5.53	2.64	2.75	2.02	59.07	31.35	31.02	22.74
4.	X_4	94.20	95.39	96.55	90.69	57.86	31.03	40.27	31.06	78.24	41.19	52.04	41.11
5.	X_5	92.87	97.34	97.66	91.61	82.92	33.13	54.10	33.39	84.62	59.63	102.54	58.63
6.	X_6	91.91	93.44	90.86	84.23	0.17	0.19	0.16	0.11	50.75	51.67	47.44	32.19
7.	X_7	55.49	82.27	46.28	33.17	3.85	5.76	1.46	1.27	6.99	3.19	2.92	2.38
8.	X_8	96.73	94.89	82.79	82.23	0.35	0.37	0.33	0.22	11.15	11.95	11.06	7.31
9.	X_9	92.82	86.80	56.35	45.69	0.17	0.21	0.15	0.07	9.43	12.68	9.28	4.33
10.	X_{10}	95.44	90.12	87.74	72.41	0.81	0.65	1.30	0.47	28.12	20.97	33.96	14.48
11.	X_{11}	99.28	70.09	98.97	97.99	4.43	0.59	4.56	3.02	64.69	12.37	64.48	48.48
12.	X_{12}	98.09	72.60	92.98	86.54	0.30	0.13	0.33	0.19	64.83	19.34	57.22	33.28
13.	X_{13}	94.93	96.63	97.75	93.37	5.28	5.72	6.61	4.07	12.19	13.50	15.78	9.57
14.	X_{14}	79.13	67.13	96.86	78.41	3.14	2.63	6.85	3.06	71.87	52.57	122.34	61.41
15.	X_{15}	71.77	69.51	93.40	67.27	9.51	8.82	15.20	7.32	46.65	41.14	72.62	34.97
16.	X_{16}	86.01	87.38	90.96	78.79	12.83	15.77	16.56	9.98	58.87	65.61	65.33	42.08

(Source: Thirugnanakumar, 1991).

Chapter 2

Correlation and Causation

Seed yield is the foremost important attribute and is the end product of many complex component characters which singly or jointly influence the seed yield. Seed yield does not possess gene *per se* as such. It is the interaction phenotype of the harmonious understanding, mutual adjustment and manifestation of its component characters. As most of the component characters are polygenic, they are greatly influenced by the environment. So, the selection of genotypes based on yield as such is not likely to be effective. The information on the strength and direction of association of the component characters with seed yield and also among themselves will be very useful in formulating an effective breeding programme for seed yield improvement in a given time. Correlation studies provide about the magnitude and direction of genetic association between the seed yield and its component characters.

Correlation coefficients provide the measure of association between two characters. But, they do not provide the causal basis of such association. Further, correlation between seed yield and its component characters are often misleading due to the inter-relationship that exist among the component characters and the operation of size symmetry and component compensation that thwart the breeding progress. The direct contribution of each component to the seed yield and the indirect effects it has through its association with other components cannot be differentiated from mere correlation studies. This is how, it became necessary for a plant breeder to have information on the direct and indirect effects of these components on the seed yield to identify the key characters for improvement.

Wright (1921) developed a technique, the path-coefficient analysis, which is an effective biometric tool for understanding the direct and indirect effects of the component characters on seed yield. It also permits the critical examination of specific factors that produce a given correlation. Thus, correlation and path-coefficient analysis assume special importance in deciding the basis for selection of desired characters and the direction through which seed yield improvement can be made.

It is of prime importance to recognize the nature and size of the population under study, in as much as the magnitude of the correlation coefficient can often be influenced by the choice of individuals upon which the observations are made (Dewey and Lu, 1959). Hence, a large number of diverse genotypes from different ecogeographic regions should be included in the study of correlation and causation. Clark (1987) has theoretically shown that genetic correlations do depend upon the environment and that genotype-by-environment interactions affect genetic correlations. Hence, the study of correlation and causation should be conducted over environments or season or location.

2.1. Correlation studies

Phenotypic, genotypic and environmental correlation coefficients are calculated among mean values of seed yield and its component characters using the method given by Johnston *et al.* (1955).

a) Phenotypic correlation coefficients = r_p (1.2)

$$r_p(1.2) = \frac{\text{Phenotypic covariance between the characters 1 and 2}}{(\text{Phenotypic variance of character 1})^{1/2} \times (\text{Phenotypic variance of character 2})^{1/2}}$$

b) Genotypic correlation coefficients = r_g (1.2)

$$r_g(1.2) = \frac{\text{Genotypic covariance between the characters 1 and 2}}{(\text{Genotypic variance of character 1})^{1/2} \times (\text{Genotypic variance of character 2})^{1/2}}$$

c) Environmental correlation coefficients = r_e(1.2)

$$r_e(1.2) = \frac{\text{Environmental covariance between characters 1 and 2}}{(\text{Environmental variance of character 1})^{1/2} \times (\text{Environmental variance of character 2})^{1/2}}$$

2.1.1. Example

The phenotypic, genotypic and environmental correlation coefficients were worked out by Thirugnanakumar (1991) by utilizing the data from 60 genotypes of sesame, sown in three replications and studied over three seasons for 16 seed yield and its contributing traits (Tables 8, 9, 10 and 11). The genotypic correlation coefficients were mostly slightly higher than the phenotypic correlation coefficients, especially in almost all the cases wherever the correlation coefficients are significant. This may be due to the effect of environment in modifying the total expression of the genotypes, thus altering the phenotypic expression (Nandpuri *et al.*, 1973).

The character association analysis from the Tables 8, 9, 10 and 11 revealed that the traits *viz.*, number of flowers (X_4), number of capsules (X_5), weight of 1000 seeds (X_{10}) and harvest index (X_{16}) were positively and significantly associated with seed yield consistently in all the three seasons and pooled analysis. This indicated the high reliability of these traits in the selection programme. The traits *viz.*, plant height at maturity (X_2), number of branches (X_3), seed density (X_{12}) and total dry matter production were positively and significantly associated with seed yield mostly (in three analyses). It indicated the seasonal influence on the character association. Days to maturity (X_1) and number of seeds per capsule (X_7) were positively and significantly associated with seed yield in only one season, showing the vulnerability of association to the seasonal changes to a greater extent.

The information on the association of seed yield with other traits may not alone be sufficient as inter-relationships between the traits may also affect the overall relationship of the characters. Doku (1970) suggested that the selection based on yield components would be effective in improving the yield, provided the components are highly heritable and the genotypic correlations among them are not negative. Negative genotypic correlation between the components characters may hinder the rate of improvement of other desirable characters and nullify the progress. The correlation coefficients in the Tables 8, 9, 10 and 11 revealed that the intercorrelation among the traits which exhibited constant significant positive correlation with seed yield *viz.*, plant height (X_2), number of branches (X_3), seed density (X_{12}) and total dry matter production (X_{15}) were mostly positive. Hence, selection and manipulation of these traits would not affect each other. Similarly, the inter-correlation among seed yield and the component characters which were consistently (in any three analyses) had significant

Table 8 : Phenotypic, genotypic and environmental correlation coefficients – S I

		X_2	X_3	X_4	X_5	X_6	X_7	X_8	X_9	X_{10}	X_{11}	X_{12}	X_{13}	X_{14}	X_{15}	X_{16}
X_1	P	0.0896	0.1571	0.0794	0.1198	0.1575	0.0583	0.1426	0.1174	0.2315	0.1695	-0.1072	0.0257	0.0879	0.0840	0.0177
	G	0.4104**	0.1666	0.0752	0.1181	0.1620	0.0951	0.1480	0.1241	0.2576*	0.1788	-0.1082	0.0434	0.0907	0.0996	0.0175
	E	-0.0367	0.0324	0.1447	0.1427	0.1011	-0.0605	0.0383	0.0253	-0.2199	-0.1286	-0.0959	-0.2673*	0.0848	0.0175	0.0202
X_2	P		0.0547	0.0673	0.0493	0.0365	-0.0503	0.0769	-0.0315	0.0679	-0.0862	-0.0491	0.0425	-0.0440	0.1280	-0.1799
	G		0.3722**	0.4639**	0.3996**	-0.0322	0.2723*	0.2821*	-0.1945	0.4968**	0.2919*	-0.0985	0.2314	-0.2376	1.2864**	-0.5620**
	E		-0.1265	-0.1910	-0.1792	0.1602	-0.1557	0.0457	0.0580	-0.2544*	0.1699	-0.1858	-0.0620	-0.2180	-0.2771*	-0.1393
X_3	P			0.3190*	0.2792*	-0.0846	-0.2494	0.0011	-0.0815	0.2569*	-0.0706	0.1935	-0.1206	0.2239	0.2517	0.0301
	G			0.3266*	0.2876*	-0.0909	-0.3569*	-0.0008	-0.0823	0.2655*	-0.0780	0.2023	-0.1202	0.2292	0.2423	0.0370
	E			0.2181	0.1749	-0.0116	0.0323	0.0365	-0.0715	0.1307	0.1716	0.0270	-0.1295	0.2210	0.3688**	-0.0273
X_4	P				0.9856**	0.0800	-0.1972	0.1098	0.0812	0.3273*	-0.1054	0.2620*	0.2537*	0.5066**	0.7225**	-0.0141
	G				0.9869**	0.0788	-0.3095*	0.1214	0.0947	0.3432**	-0.1090	0.2692*	0.2704*	0.5431**	0.7935**	-0.0335
	E				0.9736**	0.0973	0.1656	-0.1393	-0.1136	0.0370	0.0022	0.0972	-0.0355	0.3428**	0.5481**	-0.1220
X_5	P					0.0824	-0.2019	0.0577	0.0424	0.2823*	-0.0938	0.2271	0.2463	0.5189**	0.7270**	-0.0271
	G					0.0813	-0.3278**	0.0697	0.0531	0.2972*	-0.0990	0.2360	-0.2634*	0.5511**	0.7933**	-0.0180
	E					0.0954	0.1878	-0.1714	-0.0959	0.0439	0.0542	0.0515	-0.0175	0.3812**	0.5592**	-0.1103
X_6	P						0.4524**	0.1633	0.1378	-0.0535	0.2816*	-0.2253	0.1328	-0.0757	0.0418	-0.1117
	G						0.5056**	0.1728	0.1530	-0.0596	0.2931*	-0.2396	0.1611	-0.1040	0.0311	-0.1404
	E						0.4814**	0.0069	-0.0456	0.0374	0.0652	0.0555	-0.2765*	0.1001	0.1091	0.1232
X_7	P							0.1460	0.1820	-0.1656	0.0983	-0.1421	0.0003	-0.1559	-0.1158	-0.0420
	G							0.1826	0.1900	-0.2257	0.1310	-0.2030	0.0190	-0.3123*	-0.3137*	-0.1013
	E							0.1012	0.2564*	-0.0092	0.0191	0.0834	-0.0710	0.1674	0.2319	0.1121
X_8	P								0.6495**	0.5048**	0.0532	0.2182	0.2525*	-0.0296	0.1421	-0.1885
	G								0.6622**	0.5165**	0.0528	0.2233	0.2718*	-0.0366	0.1823	-0.2330
	E								0.4549**	0.2197	0.0935	0.0295	-0.1940	0.0290	-0.1016	0.2199
X_9	P									0.2570*	-0.0415	0.1430	0.2776*	-0.0649	0.0623	-0.1952
	G									0.2698*	-0.0479	0.1546	0.2904*	-0.0886	0.0706	-0.2251
	E									0.0529	0.1992	-0.1223	0.0826	0.0898	0.0328	0.0590

X_{10}	P	-0.1734	- 0.5346**	0.1799	0.2465	0.3073*	0.0144
	G	-0.1806	0.5348**	0.1931	0.2769*	0.3442**	0.0227
	E	0.1362	0.5798**	-0.0819	0.0601	0.1973	-0.0771
X_{11}	P		- 0.8639**	-0.1056	-0.0436	0.0571	-0.1477
	G		- 0.8683**	-0.1105	-0.0595	0.0665	-0.1656
	E		- 0.6026**	0.0854	0.2351	0.0224	0.1668
X_{12}	P			0.1881	0.1753	0.1198	0.1401
	G			0.1991	0.2058	0.1231	0.1675
	E			-0.1299	-0.0952	0.2246	-0.2660*
X_{13}	P				0.1753	0.1648	0.0676
	G				0.2023	0.2071	0.0697
	E				0.0003	-0.0521	0.0544
X_{14}	P					0.5875**	0.6671**
	G					0.5671**	0.7108**
	E					0.6599**	0.4723**
X_{15}	P						-0.1690
	G						-0.1575
	E						-0.2276

* Significant at P = 5% ** Significant at P = 1% (Source: Thirugnanakumar, 1991).

Table 9 : Phenotypic, genotypic and environmental correlation coefficients – S II

		X_2	X_3	X_4	X_5	X_6	X_7	X_8	X_9	X_{10}	X_{11}	X_{12}	X_{13}	X_{14}	X_{15}	X_{16}
X_1	P	0.3758**	0.2740*	0.1657	0.3619**	-0.1085	-0.0042	0.1098	-0.0082	0.1995	0.1114	0.1004	0.3692**	0.2557*	0.0956	0.2269
	G	0.3947**	0.3217*	0.1715	0.3759**	-0.1285	-0.0037	0.1212	-0.0034	0.1975	0.1223	0.1114	0.3819**	0.2740*	0.0790	0.2339
	E	0.1347	0.0654	0.0530	0.0201	0.2135	0.0098	0.0991	0.0617	0.2366	0.0936	0.0670	0.0879	0.2840*	0.2504	0.1730
X_2	P		0.2945*	0.3834**	0.4912**	0.0009	0.1296	0.1848	-0.1133	0.2503	-0.1706	0.3305**	0.4081**	0.4312**	0.2967*	0.2275
	G		0.3419**	0.4043**	0.5249**	0.0206	0.1498	0.1958	-0.1556	0.2612*	-0.2399	0.4157**	0.4431**	0.4999**	0.3579**	0.2090
	E		0.0175	0.1053	-0.0570	-0.2357	-0.0002	0.0430	0.2287	0.1459	0.1277	-0.0464	-0.1346	0.2354	0.0772	0.3858**
X_3	P			0.2923*	0.228	0.0037	-0.1534	-0.0508	-0.2192	0.1683	-0.0792	0.1882	0.1388	0.1413	0.0603	0.1184
	G			0.3524**	0.2603*	0.0188	-0.1945	-0.0545	-0.2379	0.2192	-0.1410	0.2961*	0.1504	0.2542*	0.1094	0.1792
	E			-0.1672	-0.1133	0.1749	0.0248	-0.0323	-0.1281	-0.1298	0.1095	-0.1628	0.0780	-0.1777	-0.0873	-0.2005
X_4	P				0.6387**	-0.1123	-0.0284	-0.0964	-0.1834	0.1979	-0.1665	0.2748*	0.3609**	0.3993**	0.3282**	0.1647
	G				0.6570**	-0.1184	-0.0208	-0.1058	-0.2159	0.2212	-0.2073	0.3420**	0.3839**	0.4803**	0.3957**	0.1756
	E				0.1628	-0.0097	-0.1099	0.0882	0.1670	-0.1072	0.0257	-0.0869	-0.1937	0.1210	0.0502	0.0576
X_5	P					-0.1073	0.0777	0.1763	-0.0555	0.3283**	-0.0916	0.3483**	0.3758**	0.6776**	0.3097*	0.4680**
	G					-0.1137	0.0702	0.1852	-0.0625	0.3488**	-0.1192	0.4167**	0.3905**	0.8310**	0.3903**	0.4970**
	E					0.0268	0.2161	-0.0476	0.0317	0.0324	0.0771	-0.0227	-0.0963	0.0627	-0.1266	0.1664
X_6	P						0.6275**	0.1541	0.0363	0.0687	0.0493	0.0448	0.0084	-0.1225	-0.0411	-0.0340
	G						0.7299**	0.1642	0.0670	0.0728	0.0349	0.0666	0.0055	-0.1318	-0.0533	-0.0165
	E						-0.164	-0.0087	-0.2580*	0.0234	0.1502	-0.0749	0.0672	-0.1231	0.0133	-0.2102
X_7	P							0.1278	0.0606	0.0249	-0.0068	0.0466	-0.0028	0.0304	-0.0744	0.1334
	G							0.1198	0.0535	0.0309	-0.0067	0.1039	0.0157	0.0562	-0.0770	0.1605
	E							0.2307	0.1011	-0.0126	0.1912	-0.1528	-0.2174	-0.0473	-0.0697	-0.0184
X_8	P								0.4390**	0.2303	0.2711**	0.0182	0.0087	0.2694*	-0.1665	0.4023**
	G								0.4475**	0.2470	0.3072*	0.0406	0.0143	0.3251*	-0.2065	0.4335**
	E								0.4004**	0.0279	0.1663	-0.1308	-0.1200	0.0769	0.0098	0.0094
X_9	P									-0.0106	0.3142*	-0.2266	-0.1145	0.0094	-0.1249	0.0909
	G									0.0014	0.3400**	-0.2230	-0.1090	-0.0382	-0.1688	0.0617
	E									-0.1041	0.2465	-0.2608*	-0.2204	0.1851	0.0311	0.2882*
X_{10}	P										0.1961	0.7399**	0.1472	0.4282**	0.0956	0.3391**
	G										0.2983*	0.7758**	0.1454	0.4814**	0.0586	0.3494**
	E										-0.2380	0.6828**	0.1999	0.2980*	0.2835*	0.2601*

X_{11}	P	-0.5076**	-0.0222	-0.1218	-0.1218	-0.0194
	G	-0.3694**	-0.0175	-0.2172	-0.2089	-0.0349
	E	-0.8528**	-0.0780	0.0866	0.0794	0.0409
X_{12}	P		0.1505	0.4508**	0.1601	0.3087*
	G		0.1580	0.6059**	0.1953	0.3631**
	E		0.1836	0.0926	0.0739	0.1048
X_{13}	P			0.2173	0.1378	0.1125
	G			0.2490	0.1549	0.1167
	E			0.1597	0.1068	0.0815
X_{14}	P				0.3827**	0.7164**
	G				0.2073	0.7643**
	E				0.7614**	0.6433**
X_{15}	P					-0.3188**
	G					-0.4356**
	E					0.1058

* Significant at P = 5%   ** Significant at P = 1%   (Source: Thirugnanakumar, 1991).

Table 10 : Phenotypic, genotypic and environmental correlation coefficients – S III

		X_2	X_3	X_4	X_5	X_6	X_7	X_8	X_9	X_{10}	X_{11}	X_{12}	X_{13}	X_{14}	X_{15}	X_{16}
X_1	P	0.0107	-0.1109	-0.1030	-0.0851	0.0316	0.1429	-0.1295	-0.1099	-0.2215	-0.0067	-0.1572	-0.2019	-0.0960	-0.1353	-0.1799
	G	0.0159	-0.1122	-0.1065	-0.0856	0.0445	0.1619	-0.1519	-0.1630	-0.2333	-0.0085	-0.1595	-0.2021	-0.1009	-0.1445	-0.1978
	E	-0.0392	-0.2056	0.0087	-0.0659	-0.2116	0.2897*	0.1035	0.1023	-0.1033	0.1026	-0.1268	-0.1965	0.0717	0.0609	0.1308
X_2	P		0.5893**	0.5588**	0.5703**	0.1617	0.0793	0.0893	0.1424	0.3170*	-0.0586	0.2001	0.2076	0.5451**	0.3725**	0.5397**
	G		0.6578**	0.6419**	0.6370**	0.1992	0.0903	0.0749	0.1925	0.3985**	-0.0798	0.2617*	0.2451	0.6218**	0.4374**	0.6316**
	E		0.4031**	0.0546	0.2396	-0.0346	0.0721	0.1481	0.0491	-0.0646	0.2293	-0.1692	-0.0720	0.0910	0.0103	0.0748
X_3	P			0.5059**	0.4562**	0.1552	0.1932	0.0051	0.1191	0.3078	0.0558	0.0946	0.2275	0.3759**	0.3531**	0.2454
	G			0.6143**	0.5523**	0.1888	0.2887*	0.0117	0.1865	0.3965**	0.0531	0.1402	0.2949*	0.4663**	0.4546**	0.3086*
	E			-0.0007	-0.0143	0.0266	0.0715	-0.0166	0.0049	-0.0183	0.2092	-0.1292	-0.2067	-0.0896	-0.1080	-0.0072
X_4	P				0.7586**	0.1171	0.2052	-0.0192	0.0863	0.3841**	-0.0194	0.2676	0.4510**	0.6486**	0.4978**	0.4984**
	G				0.7759**	0.1267	0.3792**	-0.0295	0.1194	0.4165**	-0.0173	0.2810*	0.4625**	0.6751**	0.5080**	0.5556**
	E				0.1837	-0.0269	- 0.3543**	0.0929	-0.0145	0.0104	-0.1328	0.0279	0.0604	-0.1291	0.3221*	- 0.3994**
X_5	P					0.2229	0.1803	0.0474	0.0508	0.3975**	- 0.2588*	0.4377**	0.2967*	0.8430**	0.6771**	0.7038**
	G					0.2359	0.2720*	0.0370	0.0820	0.4235**	- 0.2605*	0.4534**	0.3853*	0.8576**	0.7028**	0.7500**
	E					0.0136	-0.0228	0.2229	-0.0990	0.1034	-0.1712	0.1409	-0.0673	-0.0043	0.1478	-0.0667
X_6	P						0.4586**	0.0435	0.0247	-0.0223	-0.2315	0.1494	0.0110	0.1869	0.1840	0.1427
	G						07154**	0.0737	0.0597	-0.0029	-0.2421	0.1656	0.0100	0.1915	0.1813	0.1679
	E						-0.0238	-0.1627	-0.0905	-0.1854	-0.0627	-0.0354	0.0449	0.1345	0.2191	-0.1089
X_7	P							-0.0765	-0.0873	0.0154	-0.0937	0.0845	0.1101	0.0809	0.0490	0.0497
	G							-0.1476	-0.0702	0.0534	-0.1626	0.1895	0.1979	0.1285	0.1350	0.0055
	E							0.0490	-0.1062	-0.0726	0.2205	-0.2051	-0.2095	-0.0390	-0.2112	0.2093
X_8	P								0.0763	-0.0885	-0.0054	-0.1335	-0.0318	0.1600	0.1907	0.1699
	G								0.1698	-0.0923	-0.0085	-0.1434	-0.0127	0.1888	0.2059	0.2129
	E								-0.1447	-0.0682	0.0548	-0.0700	-0.3272*	-0.1240	0.0902	-0.1193

X_9	P	0.0162	0.1372	-0.1612	-0.0373	0.0827	0.0927	0.0797
	G	0.0082	0.1751	-0.2328	-0.0611	0.1193	0.1252	0.0905
	E	0.0451	0.0955	0.0415	0.0818	-0.0461	0.0106	0.0659
X_{10}	P		0.2223	0.3592**	0.3493**	0.3776**	0.3488**	0.2630*
	G		0.2389	0.3134*	0.3814**	0.4181**	0.3772**	0.3139*
	E		-0.0064	0.8206**	-0.0755	-0.1264	0.0820	-0.1647
X_{11}	P			-0.7644**	0.0609	-0.1895	-0.1518	-0.2168
	G			-0.7846**	0.0587	-0.1907	-0.1572	-0.2237
	E			-0.4353**	0.2092	-0.1555	-0.0231	-0.1487
X_{12}	P				0.1901	0.3520**	0.2852*	0.2983*
	G				0.2036	0.3703**	0.2980*	0.3277**
	E				-0.1019	0.0124	0.1098	-0.0382
X_{13}	P					0.1428	0.0782	0.1767
	G					0.1426	0.0759	0.1879
	E					0.1518	0.1498	-0.0104
X_{14}	P						0.8598**	0.7886**
	G						0.8819**	0.8088**
	E						0.4604**	0.5537**
X_{15}	P							0.4319**
	G							0.5024**
	E							-0.4050

* Significant at P = 5%   ** Significant at P = 1%   (Source: Thirugnanakumar, 1991).

association with seed yield *viz.*, number of flowers (X_4), number of capsules (X_5), weight of 1000 seeds (X_{10}) and harvest index (X_{16}) also showed positive and significant association among themselves. However, plant height with harvest index in SI and total dry matter production with harvest index in SII, showed strong negative association among themselves. So, selection of plant height of maturity may affect harvest index in SI and *vice-versa*. Similarly, selection of total dry matter production may affect harvest index in SII. Hence, selection of these characters has to be avoided in that seasons. The interesting learning from the Tables 8, 9, 10 and 11, is that genetic correlation coefficients were markedly different in different seasons and pooled analysis. It indicated that genetic correlation coefficients do depend on season and that genotype × season interaction affect the genetic correlation coefficients.

2.2. Path analysis

Seed yield is a complex entity and is associated with a number of other characters. Correlation is a simple measure of association of characters and does not indicate the relative contribution of causal factors to seed yield. The component characters are themselves inter-related such inter-dependence of the contributory factor often affects their direct relationship with seed yield, thereby making the correlation coefficients unreliable as selection indices.

Path coefficient analysis permits the separation of the direct effects from indirect effects through other related characters by partitioning the correlation coefficients. Accordingly to Dewey and Lu (1959), correlation simply measures the mutual association without regard to causation while the path coefficient analysis specifies the causes and measures their relative importance. Path analysis is usually done by adopting the method suggested by Dewey and Lu (1959). It is used to partition the genotypic correlation coefficients into direct and indirect effects.

2.2.1. Example

The path coefficient analyses done by Thirugnanakumar (1991) is furnished in the Tables 12, 13, 14 and 15. It is inferred from these tables, that total dry matter production (X_{15}) and harvest index (X_{16}) exerted their high positive direct effects towards seed yield, which were close to their genetic correlation coefficients with seed yield (X_{14}). Moreover, through these two traits only all the other major yield

Table 11 : Phenotypic, genotypic and environmental correlation coefficients – p

		X_2	X_3	X_4	X_5	X_6	X_7	X_8	X_9	X_{10}	X_{11}	X_{12}	X_{13}	X_{14}	X_{15}	X_{16}
X_1	P	0.0682	0.2140	0.0852	0.0918	-0.0033	0.0252	-0.0189	-0.0127	0.0373	0.1053	-0.1470	0.0033	0.0235	0.0275	-0.0615
	G	0.3114*	0.2932*	0.0864	0.0959	-0.0021	0.0207	-0.0282	-0.0368	0.0563	0.1086	-0.1568	0.0139	0.0097	0.0184	-0.0857
	E	-0.0193	-0.1111	0.0725	0.0412	-0.0140	0.0654	0.0543	0.0603	-0.0673	0.0370	-0.0635	-0.1476	0.1277	0.0886	0.1030
X_2	P		0.0667	0.0796	0.0885	0.0258	-0.0885	0.0552	0.0643	0.1613	0.0675	-0.0119	0.1413	0.0225	-0.0104	-0.0044
	G		0.4702**	0.5403**	0.5523**	-0.0303	0.1038	0.2043	0.3012*	0.9798**	0.2247	0.0526	0.6625**	0.3490**	0.4454**	0.1123
	E		-0.0459	-0.1520	-0.1405	0.0844	-0.1299	0.0252	0.0209	-0.0796	0.0987	-0.0668	-0.0457	-0.1160	-0.1781	-0.0639
X_3	P			0.4032**	0.3899**	-0.0492	-0.1212	0.0214	-0.1291	0.1846	0.0509	-0.0094	0.1953	0.2343	0.1433	0.1805
	G			0.5056	0.4884**	-0.0830	-0.2990*	0.0324	-0.2099	0.2725*	0.0460	0.0181	0.2675*	0.3318**	0.1777	0.2766*
	E			0.0658	0.0579	0.0551	0.0397	-0.0100	-0.0320	-0.0129	0.1676	-0.1081	-0.0994	-0.0158	0.0744	-0.0712
X_4	P				0.8606**	-0.0268	-0.1239	0.0647	-0.0078	0.2791*	-0.1136	0.2419	0.3848**	0.5769**	0.5790**	0.2331
	G				0.8746**	-0.0357	-0.2432	0.0732	-0.0120	0.3449**	-0.1185	0.2718*	0.4222**	0.6550**	0.6611**	0.2980*
	E				0.7177**	0.0366	0.0379	0.0113	0.0003	-0.0026	-0.0282	0.0104	-0.0470	0.1730	0.3787**	-0.1335
X_5	P					0.0128	-0.0471	0.1245	-0.0184	0.3479**	-0.2044	0.3229*	0.3139*	0.7228**	0.6387**	0.4060**
	G					0.0075	-0.1492	0.1387	-0.0102	0.4166**	-0.2153	0.3563**	0.3828**	0.8217**	0.7490**	0.4834**
	E					0.0538	0.1485	0.0337	-0.0551	0.0568	-0.0123	0.0535	-0.0420	0.1959	0.3057*	-0.0354
X_6	P						0.4260**	0.1383	0.1216	-0.0292	0.1193	-0.0850	0.1184	0.0278	0.0903	-0.0218
	G						0.6777**	0.1823	0.2504	-0.0154	0.1282	-0.0939	0.1412	0.0312	0.0912	-0.0105
	E						0.2086	-0.0802	-0.1154	-0.0827	0.0487	-0.0330	-0.0665	0.0136	0.0953	-0.0725
X_7	P							0.1074	0.0899	-0.1440	0.0343	-0.0659	0.0359	-0.0062	-0.0016	0.0375
	G							0.1553	0.1913	-0.2667*	0.0396	-0.0844	0.1107	-0.0610	-0.0833	0.0016
	E							0.0762	0.0255	-0.0309	0.1012	-0.0688	-0.1222	0.0657	0.0806	0.0977
X_8	P								0.1689	0.2097	0.0538	0.0578	0.1172	0.1676	-0.0153	0.3022*
	G								0.2733*	0.2785*	0.0540	0.0823	0.1601	0.2116	-0.0239	0.3725**
	E								0.0046	-0.0233	0.0879	-0.0755	-0.2130	-0.0116	0.0105	0.0119

X_9	P	0.1119	0.1315	-0.0533	0.0132	0.0745	0.0432	0.0635
	G	0.1800	0.1748	-0.0604	0.0209	0.0946	0.0648	0.0399
	E	0.0215	0.1387	-0.0567	-0.0021	0.0522	0.0173	0.1164
X_{10}	P		0.1140	0.2919*	0.1995	0.3134*	0.2601*	0.1895
	G		0.1387	0.1940	0.2440	0.4006**	0.3136*	0.2599*
	E		-0.0375	0.7178**	-0.0081	0.0473	0.1371	-0.0282
X_{11}	P			-0.8564**	0.0236	-0.1474	-0.0075	-0.1745
	G			-0.8951	0.0220	-0.1732	-0.0126	-0.1995
	E			-0.6178**	0.0685	0.0671	-0.0331	0.0121
X_{12}	P				0.1129	0.2833*	0.1012	0.2802*
	G				0.1250	0.3377**	0.1036	0.3424**
	E				0.0054	0.0301	0.1057	-0.0153
X_{13}	P					0.1906	0.1240	0.1297
	G					0.2095	0.1476	0.1450
	E					0.0953	0.0472	0.0453
X_{14}	P						0.6948**	0.7137**
	G						0.7104**	0.7553**
	E						0.6729**	0.5607**
X_{15}	P							0.0422
	G							0.1027
	E							-0.1235

* Significant at P = 5% ** Significant at P = 1% (Source: Thirugnanakumar, 1991).

Table 12 : Path coefficients showing the direct and indirect effects – S I Weight of 1000 seeds as dependent variable

Characters	X_1	X_2	X_3	X_4	X_5	X_6	X_7	X_8	X_9	X_{11}	X_{12}	X_{13}	X_{14}	X_{15}	X_{16}
X_1	**0.0008**	0.3133	-0.0195	-0.0101	0.0846	-0.1439	-0.0027	-0.0124	0.0250	-0.0083	0.1350	-0.1233	-0.0002	0.0110	-0.0004
X_2	0.0022	**0.1286**	-0.0475	-0.0226	0.5221	-0.4867	0.0005	-0.0356	0.0477	0.0130	0.2205	-0.1121	-0.0013	0.2540	0.0140
X_3	0.0021	0.0522	**-0.0177**	-0.0608	0.3676	-0.3504	0.0015	0.0466	-0.0001	0.0055	-0.0589	0.2302	0.0007	0.0478	-0.0009
X_4	0.0050	0.0236	-0.0220	**-0.0198**	1.1254	-1.2021	-0.0013	0.0404	0.0205	-0.0063	-0.0823	0.3063	-0.0016	0.1567	0.0008
X_5	0.0051	0.0370	-0.0190	-0.0175	**1.1106**	-1.2181	-0.0013	0.0428	0.0118	-0.0035	-0.0747	0.2685	-0.0015	0.1566	-0.0004
X_6	-0.0010	0.0508	0.0015	0.0055	0.0887	**-0.0991**	-0.0164	-0.0661	0.0292	-0.0102	0.2214	-0.2727	-0.0009	0.0061	0.0035
X_7	-0.0029	0.0298	-0.0129	0.0217	-0.3484	0.3993	**-0.0083**	-0.1307	0.0309	-0.0127	0.0989	-0.2310	-0.0001	-0.0619	0.0025
X_8	-0.0003	0.0464	-0.0134	0.0005	0.1366	-0.0848	-0.0028	**-0.0239**	0.1691	-0.0442	0.0399	0.2541	-0.0016	0.0360	0.0056
X_9	-0.0008	0.0389	0.0092	0.0050	0.1066	-0.0646	-0.0025	-0.0248	**0.1120**	-0.0668	-0.0362	0.1760	-0.0017	0.0139	0.0056
X_{11}	-0.0005	0.0560	-0.0139	0.0047	-0.1227	0.1206	-0.0048	-0.0171	0.0089	**0.0032**	0.7552	-0.9881	0.0006	0.0131	0.0041
X_{12}	0.0019	-0.0340	0.0047	-0.0123	0.3029	-0.2875	0.0039	0.0265	0.0378	-0.0103	**-0.6558**	1.1380	-0.0011	0.0243	-0.0042
X_{13}	0.0019	0.0136	-0.0110	0.0073	0.3043	-0.3209	-0.0027	-0.0025	0.0460	-0.0194	-0.0835	**0.2266**	-0.0057	0.0409	-0.0017
X_{14}	0.0092	0.0284	-0.0113	-0.0139	0.6112	-0.6713	0.0017	0.0480	-0.0062	0.0059	-0.0450	0.2342	**-0.0012**	0.1120	-0.0177
X_{15}	0.0052	0.0312	-0.0611	-0.0147	0.8930	-0.9664	-0.0005	0.0410	0.0308	-0.0047	0.0502	0.1401	-0.0012	**0.1974**	0.0039
X_{16}	0.0065	0.0055	0.0267	-0.0022	-0.0377	0.0219	0.0023	0.0132	-0.0377	0.0150	-0.1251	0.1906	-0.0004	-0.0311	**-0.0249**

Residual effect = 0.5422 (Source: Thirugnanakumar, 1991).

Table 12(i) : Path coefficients showing the direct and indirect effects – S I

Characters	X_1	X_2	X_3	X_4	X_5	X_6	X_7	X_8	X_9	X_{10}	X_{11}	X_{12}	X_{13}	X_{15}	X_{16}
X_1	**0.0002**	-0.0322	0.0226	0.0020	-0.0026	0.0203	-0.0009	-0.0048	-0.0038	0.0137	0.0164	-0.0035	-0.0011	0.0494	0.0149
X_2	0.0004	**-0.0132**	0.0551	0.0044	-0.0158	0.0688	0.0002	-0.0139	-0.0072	-0.0215	0.0268	-0.0032	-0.0059	0.6386	-0.4760
X_3	0.0002	-0.0054	**0.0205**	0.0118	-0.0111	0.0495	0.0005	0.0182	0.0002	-0.0091	-0.0072	0.0065	0.0031	0.1203	0.0314
X_4	0.0003	-0.0024	0.0256	**0.0039**	-0.0341	0.1699	-0.0005	0.0158	-0.0031	0.0105	-0.0100	0.0087	-0.0069	0.3938	-0.0283
X_5	0.0002	-0.0038	0.0220	0.0034	**-0.0337**	0.1722	-0.0005	0.0167	-0.0018	0.0059	-0.0091	0.0076	-0.0067	0.3938	-0.0152
X_6	-0.0004	-0.0052	-0.0018	-0.0011	-0.0027	**0.0140**	-0.0057	-0.0257	-0.0044	0.0169	0.0269	-0.0077	-0.0041	0.0155	-0.1189
X_7	-0.0002	-0.0031	0.0150	-0.0042	0.0106	-0.0564	**-0.0029**	-0.0509	-0.0046	0.0210	0.0120	-0.0065	-0.0005	-0.1557	-0.0858
X_8	0.0004	-0.0048	0.0155	-0.0001	-0.0041	0.0120	-0.0010	**-0.0093**	-0.0254	0.0733	0.0049	0.0072	-0.0069	0.0905	-0.1888
X_9	0.0002	-0.0040	-0.0107	-0.0010	-0.0032	0.0091	-0.0009	-0.0097	**-0.0168**	0.1107	-0.0044	0.0050	-0.0074	0.0351	-0.1907
X_{10}	0.0007	-0.0083	0.0274	0.0031	-0.0017	0.0512	0.0003	0.0115	-0.0131	**0.02999**	-0.0166	0.0172	-0.0049	0.1709	0.0193
X_{11}	-0.0001	-0.0057	0.0161	-0.0009	0.0037	-0.0170	-0.0017	-0.0067	-0.0013	-0.0053	**0.0919**	-0.0280	0.0024	0.0330	-0.1402
X_{12}	0.0004	0.0035	-0.0054	0.0024	-0.0092	0.0406	0.0014	0.0103	-0.0057	0.0171	-0.0798	**0.0322**	-0.0051	0.0611	0.1419
X_{13}	0.0001	-0.0014	0.0128	-0.0014	-0.0092	0.0454	-0.0009	-0.0010	-0.0069	0.0322	-0.0102	0.0064	**-0.0255**	0.1028	0.0591
X_{15}	0.0003	-0.0032	0.0709	0.0029	-0.0271	0.1366	-0.0002	0.0160	-0.0046	0.0078	0.0061	0.0040	-0.0053	**0.4964**	-0.1334
X_{16}	0.0002	-0.0006	-0.310	0.0004	0.0011	-0.0031	0.0008	0.0052	0.0047	-0.0249	-0.0152	0.0054	-0.0018	-0.0782	**0.8469**

Residual effect = 0.1534 (Source: Thirugnanakumar, 1991).

Note: Figures bold indicate direct effects.

Table 13 : Path coefficients showing the direct and indirect effects – S II Weight of 1000 seeds as dependent variable

Characters	X_1	X_2	X_3	X_4	X_5	X_6	X_7	X_8	X_9	X_{11}	X_{12}	X_{13}	X_{14}	X_{15}	X_{16}
X_1	**0.0614**	0.0173	-0.0014	0.0007	0.0044	-0.0163	-0.0042	0.0001	-0.0012	0.0001	0.0833	0.1105	-0.0058	-0.0089	-0.0424
X_2	0.1121	**0.0068**	-0.0035	0.0007	0.0104	-0.0282	0.0007	-0.0054	-0.0020	0.0004	-0.1634	0.4121	-0.0067	-0.0404	-0.0379
X_3	0.0570	0.0056	**-0.0012**	0.0021	0.0091	-0.0113	-0.0006	0.0071	0.0006	0.0007	-0.0961	0.2936	-0.0023	-0.0124	-0.0325
X_4	0.1077	0.0030	-0.0014	**0.0007**	0.0257	-0.0286	-0.0038	0.0008	0.0011	0.0006	-0.1412	0.3391	-0.0058	-0.0448	-0.0319
X_5	0.1863	0.0065	-0.0018	0.0005	**0.0169**	-0.0435	-0.0037	-0.0025	-0.0019	0.0002	-0.0812	0.4132	-0.0059	-0.0442	-0.0902
X_6	-0.0296	-0.0022	-0.0007	-0.0004	-0.0030	**0.0049**	0.0324	-0.0265	-0.0017	-0.0002	0.0238	0.0660	-0.0008	0.0060	0.0030
X_7	0.0126	-0.0006	-0.0005	-0.0004	-0.0005	-0.0031	**0.0236**	-0.0363	-0.0012	-0.0002	-0.0456	0.1030	-0.0002	0.0087	-0.0291
X_8	0.0729	0.0021	-0.0007	-0.0001	-0.0027	-0.0081	0.0053	**-0.0043**	-0.0101	-0.0013	0.2092	0.0402	-0.0002	0.0234	-0.0786
X_9	-0.0086	-0.0006	0.0005	-0.0005	-0.0056	0.0027	0.0022	-0.0019	**-0.0045**	-0.0029	0.2315	-0.2211	0.0016	0.0191	-0.0112
X_{11}	-0.0487	0.0021	0.0008	-0.0003	-0.0053	0.0052	0.0011	0.0024	-0.0031	**-0.0010**	0.6810	-0.3663	0.0003	0.0236	0.0063
X_{12}	0.1358	0.0019	-0.0014	0.0006	0.0088	-0.0181	0.0022	-0.0038	-0.0004	0.0006	**-0.2516**	0.9915	-0.0024	-0.0221	-0.0659
X_{13}	0.0558	0.0066	-0.0015	0.0003	0.0099	-0.0170	0.0002	-0.0006	-0.0001	0.0003	-0.0119	**0.1572**	-0.0151	-0.0175	-0.0212
X_{14}	0.2242	0.0047	-0.0017	0.0005	0.0124	-0.0361	-0.0043	-0.0020	-0.0033	0.0001	-0.1479	0.6007	**-0.0038**	-0.0235	-0.1386
X_{15}	0.0465	0.0014	-0.0012	0.0002	0.0102	-0.0170	-0.0017	0.0028	0.0021	0.0005	-0.1422	0.1937	-0.0234	**-0.1132**	0.0790
X_{16}	0.1714	0.0040	-0.0007	0.0004	0.0045	-0.0216	-0.0005	-0.0058	-0.0044	-0.0002	-0.0238	0.3600	-0.0018	0.0493	**-0.1814**

Residual effect = 0.5422 (Source: Thirugnanakumar, 1991).

Table 13(i) : Path coefficients showing the direct and indirect effects – S II

Characters	X_1	X_2	X_3	X_4	X_5	X_6	X_7	X_8	X_9	X_{10}	X_{11}	X_{12}	X_{13}	X_{15}	X_{16}
X_1	**1.2590**	-0.0851	0.0017	-0.0008	-0.0260	0.0732	0.0223	-0.0009	0.0063	-0.0002	-0.5298	-0.7114	0.0358	0.0413	0.1883
X_2	1.6650	**-0.0336**	0.0043	-0.0008	-0.0613	0.1022	-0.0036	0.0366	0.0102	-0.0007	1.0387	-2.6538	0.0415	0.1868	0.1683
X_3	1.3975	-0.0274	**0.0015**	-0.0023	-0.0535	0.0507	0.0032	-0.0475	-0.0029	-0.0011	0.6108	-1.8905	0.0141	0.0573	0.1442
X_4	1.4102	-0.0146	0.0017	**-0.0008**	-0.1517	0.1280	0.0205	-0.0051	-0.0055	-0.0010	0.8978	-2.1836	0.0360	0.2070	0.1414
X_5	2.2233	-0.0320	0.0023	-0.0006	**-0.0997**	0.1948	0.0197	0.0172	0.0097	-0.0003	0.5160	-2.6603	0.0366	0.2042	0.4001
X_6	0.4642	0.0109	0.0001	0.0004	0.0180	**-0.0221**	-0.1732	0.1784	0.0086	0.0003	-0.1512	-3.4252	0.0005	-0.0279	-0.0133
X_7	0.1968	0.0003	0.0006	0.0005	0.0032	0.0137	**-0.1264**	0.2444	0.0063	0.0002	0.2897	-0.6634	0.0015	-0.0403	0.1292
X_8	1.5740	-0.0103	0.0008	0.0001	0.0161	0.0361	-0.0284	**0.0293**	0.0523	0.0020	-1.3302	-0.2590	0.0013	-0.1080	0.3490
X_9	0.0091	0.0003	-0.0007	0.0006	0.0328	-0.0122	-0.0116	0.0131	**0.0234**	0.0045	-1.4713	1.4238	-0.0102	-0.0883	0.0497
X_{10}	6.3745	-0.0168	0.0011	-0.0005	-0.0336	0.0679	-0.0126	0.0075	0.0129	**0.0001**	-1.2917	-4.9531	0.0136	0.0307	0.2813
X_{11}	1.9014	-0.0104	-0.0010	0.0003	0.0315	-0.0232	-0.0060	-0.0163	0.0161	0.0015	**-4.3302**	2.3583	-0.0016	-0.1093	-0.0281
X_{12}	4.9454	-0.0095	0.0018	-0.0007	-0.0519	0.0812	-0.0115	0.0254	0.0021	-0.0010	1.5996	**-6.3843**	0.0149	0.1022	0.2923
X_{13}	0.9269	-0.0325	0.0019	-0.0004	-0.0582	0.0761	-0.0009	0.0038	0.0007	-0.0005	0.0756	-1.0123	**0.0937**	0.0811	0.0939
X_{15}	0.3737	-0.0067	0.0015	-0.0003	-0.0600	0.0760	0.0092	-0.0188	-0.0108	-0.0008	0.9044	-1.2471	0.0145	**0.5232**	-0.3507
X_{16}	2.2274	-0.0199	0.0009	-0.0004	-0.0266	0.0968	0.0029	0.0392	0.0227	0.0003	0.1512	-2.3181	0.0109	-0.2279	**0.8051**

Residual effect = 0.1534 (Source: Thirugnanakumar, 1991).

Note: Figures bold indicate direct effects.

Table 14 : Path coefficients showing the direct and indirect effects – S III Weight of 1000 seeds as dependent variable

Characters	X_1	X_2	X_3	X_4	X_5	X_6	X_7	X_8	X_9	X_{11}	X_{12}	X_{13}	X_{14}	X_{15}	X_{16}
X_1	**0.0773**	0.1150	0.0002	-0.0098	0.0110	-0.0080	0.0020	-0.0037	-0.0045	-0.0092	-0.0114	-0.2109	-0.0069	-0.0831	-0.0912
X_2	-0.4762	**0.0018**	0.0121	0.0576	-0.0666	0.0595	0.0089	-0.0021	0.0022	0.0109	-0.1069	0.3460	0.0084	0.2516	0.2910
X_3	-0.3571	-0.0129	**0.0080**	0.0875	-0.0637	0.0516	0.0084	-0.0066	0.0003	0.0106	0.0711	0.1854	0.0101	0.2615	0.1422
X_4	-0.5170	-0.0122	0.0078	**0.0538**	-0.1037	0.0725	0.0057	-0.0086	-0.0009	0.0068	-0.0231	0.3716	0.0159	0.2922	0.2560
X_5	-0.6568	-0.0098	0.0077	0.0483	**-0.0805**	0.0934	0.0105	-0.0062	0.0011	0.0046	-0.3488	0.5995	0.0105	0.4043	0.3456
X_6	-0.1467	0.0051	0.0024	0.0165	-0.0131	**0.0220**	0.0447	-0.0162	0.0022	0.0034	-0.3242	0.2190	0.0003	0.1043	0.0774
X_7	-0.0984	0.0186	0.0011	0.0253	-0.0393	0.0254	**0.0320**	-0.227	-0.0044	-0.0040	-0.2178	0.2506	0.0068	0.0777	0.0025
X_8	-0.1446	-0.0175	0.0009	0.0010	0.0031	0.0035	0.0033	**0.0033**	0.0299	0.0096	-0.0114	-0.1896	-0.0004	0.1184	0.0981
X_9	-0.0913	-0.0187	0.0023	0.0163	-0.0124	0.0077	0.0027	0.0016	**0.0051**	0.0566	0.2345	-0.3078	-0.0021	0.0720	0.0417
X_{11}	0.1461	-0.0010	-0.0010	0.0046	0.0018	-0.0243	-0.0108	0.0037	-0.0003	**0.0099**	1.3391	-1.0375	0.0020	-0.0904	-0.1031
X_{12}	-0.2836	-0.0183	0.0032	0.0123	-0.0292	0.0423	0.0074	-0.0043	-0.0043	-0.0132	**-1.0507**	1.3293	0.0070	0.1714	0.1510
X_{13}	-0.1092	-0.0232	0.0030	0.0258	-0.0480	0.0285	0.0004	-0.0045	-0.0004	-0.0035	0.0786	**0.2692**	0.0344	0.0436	0.0866
X_{14}	-0.7659	-0.0116	0.0075	0.0408	-0.0700	0.0801	0.0086	-0.0029	0.0057	0.0068	-0.2554	0.4896	**0.0049**	0.5073	0.3727
X_{15}	-0.6574	-0.0166	0.0053	0.0398	-0.0527	0.0656	0.0081	-0.0031	0.0062	0.0071	-0.2105	0.3941	0.0026	**0.5753**	0.2315
X_{16}	-0.6194	-0.0228	0.0076	0.0270	-0.576	0.0701	0.0075	-0.0001	0.0064	0.0051	-0.2996	0.4334	0.0065	0.2890	**0.4608**

Residual effect = 0.5422 (Source: Thirugnanakumar, 1991).

Table 14(i) : Path coefficients showing the direct and indirect effects – S III

Characters	X_1	X_2	X_3	X_4	X_5	X_6	X_7	X_8	X_9	X_{10}	X_{11}	X_{12}	X_{13}	X_{15}	X_{16}
X_1	**0.0156**	0.1152	-0.0008	-0.0002	-0.0163	-0.0017	0.0014	-0.0077	-0.0012	-0.0031	-0.0017	-0.0346	0.0102	-0.0849	-0.0913
X_2	-0.0267	**0.0018**	-0.0488	0.0010	0.0983	0.0128	0.0063	-0.0043	0.0060	0.0037	-0.0156	0.0567	-0.0124	0.2569	0.2914
X_3	-0.0266	-0.0129	**-0.0321**	0.0015	0.0941	0.0111	0.0060	-0.0137	0.0001	0.0035	0.0104	0.0304	-0.0149	0.2670	0.1424
X_4	-0.0279	-0.0123	-0.0313	**0.0009**	0.1531	0.0155	0.0040	-0.0180	-0.0002	0.0023	-0.0034	0.0609	-0.0234	0.2984	0.2564
X_5	-0.0284	-0.0099	-0.0311	0.0008	**0.1188**	0.0200	0.0075	-0.0129	0.0003.	0.0016	-0.0509	0.0983	-0.0154	0.4129	0.3461
X_6	0.0002	0.0051	-0.0097	0.0003	0.0194	**0.0047**	0.0317	-0.0339	0.0006	0.0011	-0.0473	0.0359	-0.0005	0.1065	0.0775
X_7	-0.0036	0.0187	-0.0044	0.0004	0.0581	0.0054	**0.0226**	-0.0475	-0.0012	-0.0013	-0.0318	0.0411	-0.0100	0.0793	0.0026
X_8	0.0062	-0.0175	-0.0037	0.0002	-0.0045	0.0007	0.0023	**0.0070**	0.0079	0.0032	-0.0017	-0.0311	0.0006	0.1209	0.0982
X_9	-0.0005	-0.0188	-0.0094	0.0003	0.0183	0.0016	0.0019	0.0033	**0.0013**	0.0190	0.0342	-0.0505	0.0031	0.0736	0.0418
X_{10}	-0.0670	-0.0269	-0.0194	0.0006	0.0638	0.0085	-0.0009	-0.0025	-0.0007	**0.0002**	0.0467	0.0679	-0.0193	0.2216	0.1448
X_{11}	-0.0160	-0.0010	0.0039	0.0001	-0.0265	-0.0052	-0.0077	0.0077	-0.0001	0.0033	**0.1954**	-0.1701	-0.0030	-0.0924	-0.1032
X_{12}	-0.0210	-0.0184	-0.0128	0.0002	0.0430	0.0091	0.0052	-0.0090	-0.0011	-0.0044	-0.1533	**0.2168**	-0.0103	0.1750	0.1512
X_{13}	-0.0255	-0.0233	-0.0120	0.0004	0.0708	0.0061	0.0003	-0.0094	-0.0001	-0.0012	0.0115	0.0441	**-0.0505**	0.0446	0.0867
X_{15}	-0.0253	-0.0166	-0.0213	-0.0007	0.0778	0.0141	0.0057	-0.0064	0.0016	0.0024	-0.0307	0.0646	-0.0038	**0.5874**	0.2318
X_{16}	-0.0210	-0.0228	-0.0308	0.0005	0.0851	0.0150	0.0053	-0.0003	0.0017	0.0017	-0.0437	0.0710	-0.0095	0.2951	**0.4614**

Residual effect = 0.1534 (Source: Thirugnanakumar, 1991).

Note: Figures bold indicate direct effects.

Table 15 : Path coefficients showing the direct and indirect effects – Pooled, Weight of 1000 seeds as dependent variable

Characters	X_1	X_2	X_3	X_4	X_5	X_6	X_7	X_8	X_9	X_{11}	X_{12}	X_{13}	X_{14}	X_{15}	X_{16}
X_1	**0.6413**	-0.4110	0.2440	-0.0370	0.0317	0.0019	0.0133	0.0022	-0.0245	0.3280	-0.5128	0.0081	-0.0475	0.0672	-0.2485
X_2	0.1997	**-1.3198**	0.3914	-0.2312	0.1824	0.0273	0.0665	-0.0163	0.2005	0.6785	0.1721	0.3807	-1.7011	1.6236	0.3255
X_3	0.1880	-0.6205	**0.8324**	-0.2163	0.1613	0.0748	-0.1915	-0.0026	-0.1397	0.1390	0.0592	0.1560	-1.6174	0.6479	0.8019
X_4	0.0554	-0.7130	0.4209	**-0.4278**	0.2889	0.0322	-0.1558	-0.0058	-0.0080	-0.3580	0.8885	0.2463	-3.1926	2.4101	0.8638
X_5	0.0615	-0.7290	0.4066	-0.3742	**0.3303**	-0.0068	-0.0956	-0.0111	-0.0068	-0.6500	1.1650	0.2000	-4.0053	2.7306	1.4013
X_6	-0.0013	0.0399	-0.0691	0.0153	0.0025	**-0.9015**	0.4341	-0.0145	0.1667	0.3873	-0.3069	0.0824	-0.1520	0.3323	-0.0305
X_7	0.0133	-0.1370	-0.2489	0.1040	-0.0493	-0.6110	**0.6405**	-0.0124	0.1273	0.1197	-0.2761	0.0646	0.2972	-0.3035	0.0048
X_8	-0.0181	-0.2696	0.0270	-0.0313	0.0458	-0.1644	0.0995	**-0.0798**	0.1819	0.1632	0.2692	0.0934	-1.0311	-0.0872	1.0800
X_9	-0.0236	-0.3975	-0.1747	0.0051	-0.0034	-0.2257	0.1225	-0.0218	**0.6656**	0.5278	-0.1975	0.0122	-0.4612	0.2364	0.1157
X_{11}	0.0697	-0.2965	0.0383	0.0507	-0.0711	-0.1156	0.0254	-0.0043	0.1163	**3.0197**	-2.9266	0.0129	0.8440	-0.0458	-0.5783
X_{12}	-0.1006	-0.0695	0.0151	-0.1163	0.1177	0.0846	-0.0541	-0.0066	-0.0402	-2.7028	**3.2697**	0.0729	-1.6461	0.3775	0.9926
X_{13}	0.0089	-0.8612	0.2226	-0.1806	0.1132	-0.1273	-0.0709	-0.0128	0.0139	0.0665	0.4088	**0.5834**	-1.0209	0.5382	0.4203
X_{14}	0.0062	-0.4606	0.2762	-0.2802	0.2714	-0.0281	-0.0391	-0.0169	0.0630	-0.5229	1.1042	0.1222	**-4.8741**	2.5896	2.1895
X_{15}	0.0118	-0.5878	0.1479	-0.2829	0.2474	-0.0822	-0.0533	0.0019	0.0432	-0.0379	0.3386	0.0861	-3.4624	**3.6450**	0.2977
X_{16}	-0.0550	-0.1482	0.2303	-0.1275	0.1597	0.0095	0.0011	-0.0297	0.0266	-0.6024	1.1196	0.0846	-3.6816	0.3744	**2.8987**

Residual effect = 0.9799 (Source: Thirugnanakumar, 1991).

Table 15(i) : Path coefficients showing the direct and indirect effects – Pooled

Characters	X_1	X_2	X_3	X_4	X_5	X_6	X_7	X_8	X_9	X_{10}	X_{11}	X_{12}	X_{13}	X_{15}	X_{16}
X_1	**0.0571**	0.0315	-0.0009	0.0049	-0.0108	0.0007	-0.0013	0.0010	-0.0009	-0.0059	0.0068	-0.0204	-0.0004	0.0120	-0.0629
X_2	0.0178	**0.1011**	-0.0015	0.0306	-0.0621	-0.0011	-0.0065	-0.0076	0.0071	-0.01026	0.0140	0.0068	-0.0188	0.2894	0.0823
X_3	0.0167	0.0475	**-0.0031**	0.0236	-0.0549	-0.0029	0.0187	-0.0012	-0.0049	-0.0285	0.0029	0.0024	-0.0077	0.1155	0.2028
X_4	0.0049	0.0546	-0.0016	**0.0566**	-0.0983	-0.0013	0.0152	-0.0027	-0.0003	-0.0361	-0.0074	0.0353	-0.0122	0.4297	0.2185
X_5	0.0055	0.0559	-0.0015	0.0495	**-0.1124**	0.0003	0.0094	-0.0051	-0.0002	-0.0436	-0.0134	0.0463	-0.0099	0.4868	0.3544
X_6	-0.0002	-0.0031	0.0003	-0.0020	-0.0008	**0.0355**	-0.0425	-0.0067	0.0059	0.0016	0.0080	-0.0122	-0.0041	0.0592	-0.0077
X_7	0.0012	0.0105	0.0009	-0.0138	0.0168	0.0240	**-0.0627**	-0.0057	0.0045	0.0279	0.0025	-0.0110	-0.0032	-0.0541	0.0012
X_8	-0.0016	0.0207	-0.0001	0.0041	-0.0156	0.0065	-0.0097	**-0.0370**	0.0064	-0.0292	0.0034	0.0107	-0.0046	-0.0155	0.2732
X_9	-0.0021	0.0304	0.0007	-0.0007	0.0011	0.0089	-0.0120	-0.0101	**0.0234**	-0.0189	0.0109	-0.0078	-0.0006	0.0421	0.0293
X_{10}	0.0032	0.0990	-0.0009	0.0195	-0.0468	-0.0005	0.0167	-0.0103	0.0042	**-0.1047**	0.0086	0.0252	-0.0070	0.2038	0.1905
X_{11}	0.0062	0.0277	-0.0001	-0.0067	0.0242	0.0045	-0.0025	-0.0020	0.0041	-0.0145	**0.0623**	-0.1162	-0.0006	-0.0082	-0.1463
X_{12}	-0.0089	0.0053	-0.0006	0.0154	-0.0401	-0.0033	0.0053	-0.0030	-0.0014	-0.0203	-0.0557	**0.1299**	-0.0036	0.0673	0.2510
X_{13}	0.0008	0.0660	-0.0008	0.0239	-0.0385	0.0050	-0.0069	-0.0059	0.0005	-0.0255	0.0014	0.0162	**-0.0288**	0.0960	0.1063
X_{15}	0.0011	0.0450	-0.0006	0.0574	-0.0842	0.0032	0.0052	0.0009	0.0015	-0.0328	-0.0008	0.0134	-0.0043	**0.6499**	0.0753
X_{16}	-0.0049	0.0114	-0.0009	0.0169	-0.0543	-0.0004	-0.0001	-0.0138	0.0009	-0.0272	-0.0124	0.0445	-0.0042	0.0667	**0.7331**

Residual effect = 0.1436 (Source: Thirugnanakumar, 1991).

Note: Figures bold indicate direct effects.

component characters exerted their indirect positive effects towards seed yield (X_{14}). Among all the traits (Tables 8, 9, 10 and 11) which were positively and significantly associated with seed yield (X_{14}), the characters *viz.,* total dry matter production (X_{15}) and harvest index (X_{16}) alone exerted consistant positive high direct effects towards seed yield (X_{14}). Among the other characters, which were positively and significantly associated with seed yield, the traits *viz.,* number of flowers (X_4), weight of 1000 seeds (X_{10}) and density of seeds (X_{12}) exerted mostly positive direct effects towards seed yield. Total number of capsules (X_5) mostly showed high negative direct effect towards seed yield (X_{14}). Among the other traits which did not exhibit significant association with seed yield (X_{14}), the trait breadth of seeds (X_9) exhibited mostly positive direct effect towards seed yield and the trait oil content (X_{13}), had mostly showed negative direct effect towards seed yield (X_{14}). Even though, the direct effects of the other traits, other than total dry matter production (X_{15}) and harvest index (X_{16}) were less positive and high negative, their over all effect towards seed yield was positive, due to their high positive indirect effects exerted mostly through total dry matter production (X_{15}) and harvest index (X_{16}). So, the traits *viz.,* total dry matter production (X_{15}) and harvest index (X_{16}) must be given primary importance in sesame seed yield improvement programme through crop breeding.

The causal factors determining the genetic correlations do depend on season and that genotype × season interactions affect the expression as well as the contribution of the causal factors towards dependent variable, which in turn reflects on the direction and magnitude of genetic correlations. The differential contribution different component traits in different seasons towards the same dependent variables were well evidenced from the differences in the magnitude and direction of the genetic correlations among themselves as well as with the dependent variable.

Chapter 3

Discriminant Function Analysis

One of the important studies in the crop improvement programme is the discriminant function analysis. This is a method of exercising selection in plants, which show the extent of which character is genetically related to yield (Goulden, 1959). The technique of discriminant function was developed by Fisher (1936). But, the application of this method for plant selection was first made by Smith (1936) in wheat. He developed an index design for the selection of plant lines in wheat using the concept of discriminant function to derive a linear equation based on observable characteristics as the best available guide to the genetic value of each line.

Stephens (1942) reported about the interaction of yield components within the plant and within the population, which serves as an index for yielding ability. According to Hazel and Lush (1943), selection for a total score or net desirability is more efficient that selection for one trait at a time. The formulae presented by them for selection index give proper weight to each trait that are more efficient than selection for one trait at a time or for several traits with an independent culling level for each trait. According to Darlington and Mather (1949) discriminant function has been defined as a linear compound of series of varieties, obtained by giving the different varieties with individual coefficients which will minimize the difference between classes relative to variation within classes of objects of which the variates are measurements. Mather (1949) reported that these functions afford to the best available means of discriminating the classes and evaluating the genotype of a plant in terms of the observed values of various characters. Pundir and Raj (1971) reported that selection index based on three characters combination *i.e.,* seeds per siliqua, 1000 seed weight and yield per plant was efficient in toria.

3.1. The analysis

Discriminant function is worked out by following the method of Goulden (1959), as follows:

3.1.1. Genetic advance

The expected genetic advance is calculated by adopting the following formula.

$$G.A. = Z/P \frac{\Sigma_i \Sigma_j a_i g_{ij} b_i}{\sqrt{\Sigma_i \Sigma_j b_i p_{ij} b_j}}$$

where,

Z/P = selection differential, the value of which is 2.06 at 5% selection intensity

$b_1, b_2, ... a_n$ = Cofficients of the corresponding characters in the discriminant function

$a_i, ... a_n$ = Assigned weight for the 'n' characters

g_{ij} = Genotypic covariance between i^{th} and j^{th} characters

p_{ij} = Phenotypic covariance between i^{th} and j^{th} characters

3.1.2. Relative efficiency of selection index

It is expressed in percentage in comparison with genetic advance by straight selection. This was calculated as per the formula given below:

$$\text{Relative efficiency} = \frac{\text{Genetic advance by selection index}}{\text{Genetic advance by straight selection}} - 1 \times 100$$

$$= \left[\frac{x}{y} - 1\right] \times 100$$

$$\textit{where, } X = \frac{\Sigma_i \Sigma_j a_i g_{ij} a_j}{\sqrt{\Sigma_i \Sigma_j a_i p_{ij} a_j}}$$

3.1.3. Selection index and selection criteria

The mathematical description of the function (I) is known as selection index.

$$(I) = b_1p_1 + b_2p_2 + \ldots\ldots\ldots\ldots\ldots b_np_n$$

where,

$b_1 \ldots\ldots b_n$ = Weightage for 1 to n characters

$p_1 \ldots\ldots p_n$ = Mean value of the 1 to n characters

Using this function, the selection criteria (or) the index value for each genotype will be determined. Finally, on the basis of these selection criteria, the parents are arranged in the order of merit. The best 10 per cent of the genotypes will be selected for further breeding programme (Singh and Chaudhary, 1979).

3.2. Example

Thirugnanakumar (2002) studied 60 genotypes of sesame under natural saline stress condition during two seasons *viz.*, Feb-May, 2001 (Navarai) and Aug-Nov. 2001 (Samba). The pH of the soil was 7.8. The EC of the soil was 3.9 dS m^{-1}. The irrigation water had neutral pH (7.54, EC = 1.4 dS m^{-1}). Out of the 60 genotypes, only 24 genotypes were able to grow well and had a population size of more than five plants per replication. These 24 genotypes alone were sown to the next season. Observations were recorded on (X_1) days to 50 per cent flowering, (X_2) plant height at maturity, (X_3) number of branches per plant, (X_4) number of capsules per plant, (X_5) number of seeds per capsule, (X_6) 1000 seed weight, (X_7) total dry matter production, (X_8) harvest index, (X_9) primary root length, (X_{10}) secondary root length, (X_{11}) tertiary root length and (X_{12}) seed yield per plant.

3.2.1. Discriminant function analysis in first season

Discriminant function analysis in first season revealed that the characters namely, seed yield per plant, plant height at maturity and number of capsules per plant has got more positive weightage. Hence, these characters may be utilized as a choice of characters for breeding for saline tolerance coupled with high seed yield in sesame. The relative efficiency of selection over direct selection was about five per plant (Table 16). Based on the selection criteria, the genotypes namely, KKS 5, IS 366 and IVTS 2000-16 could well be used as donor for cross breeding to evolve high yielding saline tolerant sesame genotypes (Table 17).

Table 16 : Discriminant function for plant selection for saline tolerant sesame genotypes – Season I

Discriminant function	G.A. from straight selection	G.A. from discriminant function	Relative
0.6518 (X_1) + 1.1560 (X_2) + 0.4094 (X_3) + 1.0131 (X_4) + 0.9403 (X_5) + 0.2008 (X_6) + 0.7095 (X_7) + 0.6789 (X_8) + 0.6077 (X_9) – 0.0602 (X_{10}) + 0.9652 (X_{11}) + 2.4812 (X_{12})	11.1556		
	32.0111		
	36.1161		
	61.3869		
	8.9463		
	45.7329		
	31.5273		
	50.6847		
	22.1349		
	36.7215		
	45.1486		
	81.2588	84.95	4.5425

(Source: Thirugnanakumar, 2002).

Table 17 : Selection criteria for saline tolerant sesame genotypes – Season I

S. No.	Genotypes	Selection criteria	Rank
1.	AVTS-2000-2	213.1537	20
2.	AVTS-2000-4	287.8994	9
3.	AVTS-2000-19	288.7709	8
4.	AVTS-2000-11	300.2086	5
5.	AVTS-2000-13	209.2111	22
6.	IVTS-2000-16	328.8954	3
7.	IVTS-2000-20	255.4438	16
8.	IVTS-2000-21	293.4978	6
9.	KKS-5	333.8056	1
10.	KS-96002	300.3783	4
11.	VS-9104	230.8159	19
12.	TNAU-165	287.253	10

13.	KS-95011	207.7659	23
14.	KKS-9	209.7088	21
15.	Si-1615	250.8484	17
16.	GUN-14	2501784	18
17.	SVPR-1	294.6365	7
18.	CO-1	26.2258	14
19.	Si-3256	286.2142	12
20.	Si-1711	286.2885	11
21.	Si-1684	258.5105	15
22.	Si-280	273.4394	13
23.	IS-366	332.7681	2
24.	IVTS-99-112	191.1614	24

(Source: Thirugnanakumar, 2002).

3.2.2. Discriminant function analysis in second season

Discriminant function analysis in the second season revealed that seed yield followed by tertiary root length, total dry matter production and plant height have got more positive weightage. Hence, these characters may be utilized as choice of the characters for breeding for saline tolerance in sesame. The relative efficiency of selection over direct selection was about 11 per cent. This indicated that selection based on discriminant function analysis may be more effective than selection for seed yield (Table 18). Based on the selection criteria, the genotypes namely, KKS 5, IVTS 2000-16 and IS 366 may be declared as potential donor for cross breeding programme to evolve high yielding tolerant sesame genotypes (Table 19).

3.2.3. Discriminant function analysis in pooled analysis

Discriminant function analysis in pooled analysis revealed that the characters namely, tertiary root length followed by harvest index and plant height, as the choice of characters, as they have got more positive weightage. The relative efficiency of selection over direct selection was about eight per cent. It indicated that selection based on discriminant function work to be more effective than direct selections (Table 20). Based on the selection criteria, the genotypes namely, KKS-5, IVTS 2000-16 and IS 366 may be utilized as potential donor for evolving saline tolerant sesame genotypes (Table 21).

Table 18 : Discriminant function for plant selection for saline tolerant sesame genotypes – Season II

Discriminant function	G.A. from straight selection	G.A. from discriminant function	Relative
0.4759 (X_1) + 1.2435 (X_2) + 0.7154 (X_3) + 0.9252 (X_4) + 0.4306 (X_5) – 0.4082 (X_6) + 1.0212 (X_7) + 0.8895 (X_8) + 0.5625 (X_9) – 0.5517 (X_{10}) + 1.7375 (X_{11}) + 1.9052 (X_{12})	11.5841		
	30.5902		
	43.6430		
	59.8380		
	7.5843		
	45.7882		
	20.8940		
	54.6034		
	23.1051		
	38.1199		
	42.0581		
	76.9472	85.2593	10.8023

(Source: Thirugnanakumar, 2002).

Table 19 : Selection criteria for saline tolerant sesame genotypes – Season II

S. No.	Genotypes	Selection criteria	Rank
1.	AVTS-2000-2	213.0257	20
2.	AVTS-2000-4	286.4107	9
3.	AVTS-2000-19	278.0149	11
4.	AVTS-2000-11	291.8462	6
5.	AVTS-2000-13	205.127	22
6.	IVTS-2000-16	326.0178	2
7.	IVTS-2000-20	244.1853	17
8.	IVTS-2000-21	288.8416	8
9.	KKS-5	331.7005	1
10.	KS-96002	289.8282	7
11.	VS-9104	225.0781	19

12.	TNAU-165	298.8586	5
13.	KS-95011	191.4528	24
14.	KKS-9	207.2388	21
15.	Si-1615	239.8121	18
16.	GUN-14	247.3055	16
17.	SVPR-1	295.7297	5
18.	CO-1	248.7416	15
19.	Si-3256	281.4300	10
20.	Si-1711	273.4537	12
21.	Si-1684	259.9198	14
22.	Si-280	270.4677	13
23.	IS-366	318.4549	3
24.	IVTS-99-112	193.0833	23

(Source: Thirugnanakumar, 2002).

Table 20 : Discriminant function for plant selection for saline tolerant sesame genotypes – pooled data

Discriminant function	G.A. from straight selection	G.A. from discriminant function	Relative efficiency
$0.5464 (X_1) + 1.2185 (X_2) + 0.4225 (X_3) + 0.9648 (X_4) + 0.7869 (X_5) + 0.0831 (X_6) + 0.9782 (X_7) + 1.3404 (X_8) + 0.4886 (X_9) - 0.3995 (X_{10}) + 1.3638 (X_{11}) + 0.2967 (X_{12})$	10.9529		
	30.9274		
	1.05544		
	6.1932		
	8.3734		
	45.9716		
	26.1099		
	52.6982		
	24.8005		
	40.5757		
	42.8532		
	78.4383	84.8716	8.2017

(Source: Thirugnanakumar, 2002).

Table 21 : Selection criteria for saline tolerant sesame genotypes – pooled data

S. No.	Genotypes	Selection criteria	Rank
1.	AVTS-2000-2	222.2471	20
2.	AVTS-2000-4	296.1319	9
3.	AVTS-2000-19	292.1075	10
4.	AVTS-2000-11	304.9974	4
5.	AVTS-2000-13	215.8619	22
6.	IVTS-2000-16	335.7853	2
7.	IVTS-2000-20	258.767	16
8.	IVTS-2000-21	299.741	8
9.	KKS-5	341.8282	1
10.	KS-96002	303.7261	5
11.	VS-9104	236.7192	19
12.	TNAU-165	301.2897	7
13.	KS-95011	208.1245	23
14.	KKS-9	216.5133	21
15.	Si-1615	254.2649	18
16.	GUN-14	256.9649	17
17.	SVPR-1	302.6064	6
18.	CO-1	264.1144	15
19.	Si-3256	292.0978	11
20.	Si-1711	289.0401	12
21.	Si-1684	268.3332	14
22.	Si-280	280.7439	13
23.	IS-366	334.5049	3
24.	IVTS-99-112	201.0179	24

(Source: Thirugnanakumar, 2002).

Chapter 4

Stability Analysis

A successfully developed new cultivar should have stable performance and broad adaptation over a range of environments, seasons and locations, in addition to high yielding potential. Evaluating performance stability and range of adaptation is become increasingly important in breeding programmes. Since, stability and adaptability are important selection criteria in breeding programmes, in depth research on stability is needed for a better estimate of crop performance and adaptability.

Genotype × season interaction is of major performance to the plant breeder in developing stable but high yielding varieties of crop plants. When new varieties are tested over a series of seasons, the relative ranking of the varieties for any given attribute is rarely the same at each season. This increases the difficulty of identifying superior stable genotypes. Comstock and Moll (1963) have shown statistically the effect of large genotype × environment interaction in reducing the progress from selection.

In order to minimize the genotype × season interaction, stratification of the seasons can be practiced. The implication in stratification is that suitable selections could be made for unfavourable and favourable seasons encountered within the sampling units proposed. Stratification of the environments has been practiced for a long time to identify suitable selection that could be made for the narrower range of environments encountered within the sampling units proposed. However, even with this refinement of technique, the interactions of genotypes with locations in a sub-region with environments encountered at the same location in different years, frequently remain too large (Allard and Bradshaw, 1964). Little is known about the environmental factors contributing to such interactions. Sprague (1966) suggested that even if such information were available, the possibility of materially reducing the genotype × environment interaction in field experiments would remain questionable.

Generally, selection is done in one season and subsequent evaluation of the selected genotypes is done in the other season. This may result in the elimination of the genotypes which are at selection disadvantages in the immediate season but may have ample selection advantages in the other season. Finlay (1963, 1971), Perkins and Jinks (1968b) and Bucio-Alanis *et al.* (1969) have demonstrated that stability index is a heritable trait in plants, which could be better exploited in crop yield improvement programme.

A reliable method of estimating stability is one pre-requisite for effective selection for increased variability. Various procedures have been used to characterize individual varieties for behaviour in varying environmental conditions. Performance tests over a series of environments, when analysed in the individual manner, give information on genotype × environment interaction, but give no measurement of stability of individual entries.

Plaisted and Peterson (1959) presented a method to characterize the stability of yield performance when several varieties were tested in a number of locations within a year. But, Eberhart and Russell (1966) pointed out that this method would call for a large number of analyses, if large number of varieties were tested. Finlay and Wilkinson (1963) proposed a simple model for measuring the varietal adaptation. They suggested that the average yield of all cultivars grown at a particular site provide a measure of that environment for use in a regression analysis of yield stability. The mean yield and the regression could provide for selection of cultivars adapted to environments with differing productivities, when, only a small fraction of the variation for genotype × environment interaction is due to heterogeneity among regression coefficients, characterization of cultivars by the coefficient was not effective (Baker, 1969; Shukla, 1972; Freeman, 1973). Later, Eberhart and Russell (1966) modified and elaborated this method based on the regression technique for measuring the stability of genotypes. This method of stability analysis was then used by many workers (Baker, 1969; Breese, 1969; Johnson *et al.*, 1968; Perkins and Jinks, 1968a; Walton, 1976; Singh and Singh, 1980; Thirugnanakumar, 1991; Anandan *et al.*, 2000). Eberhart and Russell's method was preferable because of its explicit nature (Jowett, 1972). According to Eberhart and Russell (1966), the phenotypic stability could be measured by three parameters, namely mean performance over environments, the linear regression (b) and deviation from regression $(\overline{S}^2 d)$.

4.1. Stability parameters

The linear model proposed by Eberhart and Russell (1966) was,

$$Y_{ij} = \mu_I + b_i I_j + \delta_{ij}$$

where,

Y_{ij} = Mean performance of ith genotype in j^{th} season

μ_i = Average performance of ith genotype over all seasons

b_i = Regression coefficient that measures the response of the ith genotype to varying seasons

δ_{ij} = The deviation from regression of the ith genotype at j^{th} season

I_j = Seasonal index, as the deviation of the mean of all genotypes in j^{th} season for grand mean

such that $\sum_{j=1}^{n} I_j = 0$

a) The regression coefficient for each genotype was estimated as follows:

$$b_i = \sum_{j=1}^{n} Y_{ij} I_j \Big/ \sum_{j=1}^{n} I_j^2$$

b) The mean squared deviation from linear regression $(\overline{S}^2 d)$ for each genotype was estimated as follows:

$$(\overline{S}^2 d) = \left[\sum_j d_{ij}^2 \Big/ (n\text{-}2) \right] - (S_e^2 / r)$$

$$\sum_j d_{ij}^2 = \left[\sum_j Y_{ij}^2 - \frac{Y_{i\cdot}^2}{t}\right] - \frac{\left(\sum_j Y_{ij}\, I_{\cdot j}\right)^2}{\sum_j I_j^2}$$

where,

S_e^2 = pooled error mean square

n = number of seasons (environments)

r = number of replications

t = number of genotypes

I_j = $(\overline{Y}_i - \overline{Y}..)$

4.2. Analysis of variance for phenotypic stability

The analysis of variance was suggested by Eberhart and Russell (1966) is as given under.

Table 22 : Analysis of variance (phenotypic stability)

Source	df	SS	MS
Genotypes (G)	(g-1)	$\frac{1}{n}\sum Y^2 i. - C.F.$	MS1
Seasons (Sea)	(n-1)	$\frac{1}{g}\sum Y^2 .j - C.F.$	
G × Sea	(g-1) × (n-1)	$\sum_i \sum_j Y^2 ij - \sum Y^2 i./n - \sum Y^2 .j/g + C.F.$	
Sea (linear)	1	$\frac{1}{g}\left(\sum Y_{i.j} I_j\right)^2 \Big/ \sum_j I_j^2 = A\ (say)$	
G × sea (linear)	(g-1)	$\sum_i \left[\left(\sum Y_{ij} I_j\right)^2 / \sum I_j^2\right] - A$	MS2
Pooled deviations	g (n-2)	$\sum_i \sum_j d^2 ij$	MS3
Pooled deviations due to genotype 'i'	(n-2)	$\sum_j Y_{ij}^2 - (Y_{i.})^2 \Big/ n - \left(\sum_j Y_{ij} I_j^2\right) \Big/ \sum_j I_j^2 = \sum_j d_{ij}^2$	
Pooled error	n (r-1) × (g-1)		σ_e^2

(Source: Eberhart and Russsell, 1966)

4.3. Test of significance

a. In order to test the significance of the differences among the genotype means *i.e.,* $H_0 = H_1 = H_2 = H_3 = ...= H_{n1}$ the appropriate 'F test is defined as

$$F = MS_1/MS_3$$

b. To test that the genotypes do not differ for their regression on the seasonal index, i.e., $H_0 = b_1 = b_2 .... = b_n$

$$F = MS_2/MS_3$$

c. Individual deviation from linear regression is tested as follows;

$$F = \frac{\left(\sum_j d_{ij}^2\right) \Big/ (n-2)}{S_e^2/r}$$

4.4. Simple correlation coefficients

Simple correlation coefficients may be estimated among the regression coefficients (b) of seed yield and its component characters as well as among the mean squared deviations ($\bar{S}^2d$) of seed yield and its component characters.

$$r_{(1.2)} = \frac{\text{Covariance between characters 1 and 2}}{(\text{Variance of character 1} \times \text{Variance of character 2})^{1/2}}$$

The significance of these correlation coefficients were tested by referring to the table given by Snedecor and Cochran (1961) at (n-2) degrees of freedom.

4.5. Example

Thirugnanakumar (1991) analyzed 60 genotypes of sesame in three seasons *viz.,* September 1988; January 1989 and June 1989. He recorded data on 16 seed yield and its component characters in all the three seasons. He tested the phenotypic stability of the 60 sesame genotypes for 16 traits using Eberhart and Russell's (1966) method.

4.5.1. Analysis of variance

The analysis of variance in individual season and in pooled analysis is presented in Table 3.

4.5.2. Seasonal index

The relative ranking of the three seasons revealed that SI was best for plant height of maturity (X_2), number of branches (X_3), number of capsules (X_5), length of seeds (X_8), breadth of seeds (X_9) and oil content (X_{13}) (Table 23). Season III was best for days to maturity (X_1), number of flowers (X_4), 1000 seed weight (X_{10}), 1000 seed volume (X_{11}), seed yield (X_{14}) and harvest index (X_{16}). Season II was best for capsule volume (X_6), number of seeds per capsule (X_7), seed density (X_{12}) and total dry matter production (X_{15}). Thus, the widely different seasons of the present investigation would thus make it possible for early selection of genotypes to create varieties well adapted to fluctuating seasons.

Table 23 : Seasonal index (I) and seasonal mean ($\bar{S}$) for 16 characters in sesame

Characters		Seasons			
		S I	S II	S III	Overall mean
X_1	I	3.52	1.80	-5.32	000.00
	($\bar{S}$)	106.42	104.71	97.59	102.91
X_2	I	22.99	-10.31	-12.68	000.00
	($\bar{S}$)	85.57	52.36	50.07	62.67
X_3	I	0.47	-0.47	00.00	000.00
	($\bar{S}$)	9.36	8.42	8.88	8.89
X_4	I	-1.59	-0.23	1.82	000.00
	($\bar{S}$)	73.96	75.33	77.37	75.55
X_5	I	5.58	-1.39	-4.16	000.00
	($\bar{S}$)	62.53	55.56	52.76	56.95
X_6	I	-0.02	0.03	-0.01	000.00
	($\bar{S}$)	0.33	0.37	0.33	0.34
X_7	I	1.60	1.83	-3.43	000.00
	($\bar{S}$)	55.13	55.35	50.09	53.52
X_8	I	0.04	0.02	-0.06	000.00
	($\bar{S}$)	3.10	3.09	3.00	3.06
X_9	I	0.11	-0.06	-0.05	000.00
	($\bar{S}$)	1.80	1.63	1.64	1.89

X_{10}	I	-0.38	-0.18	0.56	000.00
	$(\bar{S})$	2.89	3.09	3.83	3.27
X_{11}	I	0.61	-1.46	0.85	000.00
	$(\bar{S})$	6.84	4.77	7.08	6.23
X_{12}	I	-0.11	0.09	0.02	000.00
	$(\bar{S})$	0.46	0.65	0.58	0.56
X_{13}	I	0.80	-0.20	-0.60	000.00
	$(\bar{S})$	43.35	42.35	41.95	42.55
X_{14}	I	-0.61	0.00	0.61	000.00
	$(\bar{S})$	4.38	4.99	5.60	4.99
X_{15}	I	-0.53	0.52	0.01	000.00
	$(\bar{S})$	20.39	21.44	20.94	20.92
X_{16}	I	-1.93	0.31	1.62	000.00
	$(\bar{S})$	21.79	24.04	25.35	23.73

(Source: Thirugnanakumar, 1991)

4.5.3. Pooled analysis of variance

The analysis of variance pooled under the seasons is presented in Table 24. The mean squares due to genotypes and that due to seasons were significant for all the characters except for plant height at maturity, where the mean squares due to seasons done was significant when tested against genotype × season interaction and pooled error. The mean squares due to genotype × season interaction was also significant for all the traits when tested against pooled error, except for plant height at maturity (X_2).

The arithmetic translation of the mean squares into the estimates the genotypic component (σ^2g) and genotype × season interaction component (σ^2gl) (Table 24) indicated the relative importance of the corresponding source of variation. The computation of σ^2g: σ^2gl ratio revealed the major share of the genotype × season component rather than the genotypic component in determining the character expression for 1000 seed weight (X_{11}; 1:1.57), seed breadth (X_9; 1:3.33), capsule volume (X_6; 1:3.33), number of seeds per capsule (X_7; 1:2.99) number of branches (X_3; 1:2.77), harvest index (X_{16}; 1:1.57), total dry matter production (X_{15}; 1:1.55); oil content (X_{13}; 1:1.54), seed yield (X_{14}; 1:1.47), days to maturity (X_1; 1:1.45), number of flowers (X_4; 1:1.43) and number of capsules (X_5; 1:1.43). For length of seeds (X_8) and density of seeds (X_{12}) the share of the genotypic component was equal to that of the genotype × season component (1:10. For plant height (X_2; 1:0.93) and 1000 seed volume (X_{11}; 1:0.72), the share of the genotypic component was more than that of genotype × season component. The predominance of s^2gl compared to s^2g for most of the characters indicated the necessity for testing the breeding materials in a number of seasons. For the characters *viz.*, seed length (X_8) and seed density (X_{12}), the equal share of the s^2g and s^2gl indicated the equal importance of G × S interaction in their expression. The narrowest ratio of s^2g: s^2gl, for 1000 seed volume (X_{11}) and plant height at maturity (X_2), indicated the lesser influence of G × S interaction component. A low G × S interaction component for characters has great significance for the plant breeder. In such a case, a variety breed and tested at one of the seasons will give a similar performance in other seasons.

4.5.4. ANOVA for phenotypic stability

The G x S interaction effects were further partitioned into linear and non-linear components (Table 25). The mean squares due to regression (linear component of G × S interaction) were significant for days to maturity (X_1), plant height at maturity (X_2), number of branches (X_3), number of flowers (X_4), number of capsules (X_5), number of seeds per capsule (X_7), 1000 seed volume (X_{11}), oil content (X_{13}), seed yield (X_{14}), total dry matter production (X_{15}) and harvest index (X_{16}). The non-linear component (pooled deviations) of the G × S interactions were also significant for the aforementioned characters, except 1000 seed volume, when tested against their respective pooled error. The ratio of non-linear: linear component revealed the preponderance of linear component for all the characters except capsule volume (X_6) and total dry matter production (X_{15}). Predominant linear component suggested that the performance of the genotypes may be predicted across the seasons with great precision (Breese, 1969; Samuel *et al.*, 1970; Singh and Singh, 1980; Thirugnankumar, 1991). This response is not only predictable but perhaps is also simply inherited (Bains and Gupta, 1972; Kumar, 1975). The prediction with precision is hardly possible for total dry matter production, where the linear component was less than the non-linear component. The performance of this character depend on seasonal influence on its expression.

Table 24 : Analysis of variance for pooled data over seasons

Characters	Means Squares				Variance		
	Seasons (DF.2)	Genotypes (DF.59)	Genotypes × Seasons (DF.118)	Error (DF.354)	$\sigma^2 g$	$\sigma^2 g1$	$\sigma^2 g : \sigma^2 g1$
X_1	2634.25**	55.27**	27.09**	0.64	9.11	13.23	1:1.45
X_2	47359.69**	2511.69	2041.05**	1828.33	113.89	106.36	1:0.93
X_3	26.84**	9.46**	8.76**	0.74	1.45	4.01	1:2.77
X_4	352.13**	1526.91**	743.28**	25.67	250.21	358.81	1:1.43
X_5	3039.06**	1743.29**	846.81**	26.20	286.18	410.31	1:1.43
X_6	0.06**	0.02**	0.01**	0.001	0.003	0.01	1:3.33
X_7	1060.75**	9.19**	9.15**	2.31	1.15	3.42	1:2.97
X_8	0.38**	0.09**	0.05**	0.003	0.02	0.02	1:1.00
X_9	1.03**	0.02**	0.02**	0.003	0.003	0.01	1:3.33
X_{10}	29.72**	0.47**	0.54**	0.03	0.07	0.26	1:3.71
X_{11}	194.44**	13.18**	3.18**	0.05	2.19	1.57	1:0.72
X_{12}	1.12**	0.06**	0.03**	0.002	0.01	0.01	1:1.00
X_{13}	62.56**	25.35**	13.17**	0.30	4.18	6.44	1:1.54
X_{14}	45.19**	17.68**	09.08**	0.78	2.82	4.15	1:1.47
X_{15}	33.34**	121.40**	67.12**	9.10	18.72	29.01	1:1.55
X_{16}	388.99**	186.50**	101.60**	8.01	29.75	46.80	1:1.57

* Significant at P = 5% level ** Significant at P = 1% level (Source: Thirugnanakumar, 1991)

Table 25 : Analysis of variance for phenotypic stability

Characters	Mean Squares					
	Genotypes (DF 59)	Seasons (linear) (DF 1)	Genotypes × Season (linear) (DF 59)	Pooled deviation (DF 60)	Pooled error (DF 354)	Nonlinear : linear ratio
X_1	27.6356**	2633.583**	35.2734**	2.7280**	0.6370	1 : 12.93
X_2	1255.844**	47359.51**	1398.183**	63.5895**	1828.327	1 : 21.99
X_3	4.7278**	26.8453**	4.5328**	3.7366**	0.7438	1 : 01.21
X_4	763.4523**	352.5123**	368.3774	257.7895**	25.6709	1 : 01.43
X_5	871.6451**	3038.928**	441.6755**	318.444**	26.1983	1 : 01.39
X_6	0.0105	0.056	0.0069	0.0068	0.0006	1 : 01.02
X_7	4.5969**	1060.997**	13.3383**	5.8055**	2.3083	1 : 02.30
X_8	0.0445	0.3754	0.0298	0.0217	0.003	1 : 01.37
X_9	0.0099	1.0317	0.0202	0.0146	0.0033	1 : 01.38
X_{10}	0.2325	29.7199**	0.5112	0.0922	0.0278	1 : 05.54
X_{11}	6.5902**	94.4338**	3.1874**	0.0791	0.0450	1 : 40.30
X_{12}	0.0296	1.1234	0.0228	0.0154	0.0015	1 : 01.48
X_{13}	12.6764**	62.6775**	6.9974**	5.4838**	0.2965	1 : 01.28
X_{14}	8.8389**	45.1927**	4.8430**	2.5473**	0.7755	1 : 01.90
X_{15}	60.6978**	33.3129**	33.2765**	41.2487**	9.1042	1 : 00.81
X_{16}	93.2499**	388.9635**	53.1940**	46.8050**	8.0107	1 : 01.14

Note: Significance of Genotypes and Genotypes ´ season interaction is against pooled deviation
Significance of pooled deviation is against pooled error. (Source: Thirugnanakumar, 1991)

4.5.5. Stability for seed yield

Eberhart and Russell (1966) defined a stable genotype as one which had a high mean yield, regression coefficient (b) around unity and deviation from regression nearer to zero. Later on, Breese (1969), Samuel *et al.* (1970), Paroda and Hayes (1971), Cross (1977) and Choudhary and Paroda (1980) advocated that linear regression (b) could simply be regarded as a measure of response of particular genotype, whereas the deviation from regression $(\bar{S}^2 d)$ as a measure of stability.

Thirugnanakumar (1991) selected the genotypes which had high mean (grand mean + 2SE) from the total population of 60 genotypes and classified into four groups according to the methodology followed by Mehra and Ramanujam (1979) and Singh and Singh (1980) (Table 26). He has also made association analysis among mean, 'b's and $(\bar{S}^2 d)$'s.

Table 26 : Classification of genotypes based on mean, regression coefficient (b) and deviation from regression coefficient

Group	Mean	b	$(\bar{S}^2 d)$
I	High mean	Around unity	Around zero
II	High mean	Significantly deviating from unity	Around zero
III	High mean	Significantly deviating from unity	Significantly deviating from zero
IV	High mean	Around unity	Significantly deviating from zero

(Source: Mehra and Ramanujam, 1979; Singh and Singh, 1980)

Population falling in group I will have average responsiveness and highly stable over seasons. Populations in group II will have above or below average responsiveness, that is they will be suited for stress or favourable environments and will be stable in the respective environments. Behaviour of the population falling in group III and IV can not be predicted. The estimates of the stability parameters for seed yield is presented in Table 27.

Table 27 : Estimation of stability parameters – Seed yield (g)

Genotypes	Mean	b	$\bar{S}^2d$	Genotypes	Mean	b	$\bar{S}^2d$
G1	3.90	-0.05	6.58**	G31	9.92	-1.70	1.23**
G2	4.90	-2.31*	-0.12	G32	7.38	0.004	0.91**
G3	5.59	3.78	0.02	G33	4.86	-5.22	2.32**
G4	4.60	2.31	0.04	G34	7.38	2.66	2.78**
G5	4.79	1.61	3.51**	G35	5.71	-6.41	12.88**
G6	4.29	-1.09	4.44**	G36	5.24	-0.41	0.82*
G7	6.99	-1.72	4.74**	G37	4.84	-0.90*	-0.15
G8	3.91	1.58	12.06**	G38	6.19	2.31	4.69**
G9	4.75	1.75	9.61**	G39	6.44	3.01	3.30**
G10	4.63	3.34	7.71**	G40	5.73	-0.32	17.37**
G11	7.47	2.56	10.55**	G41	1.76	-4.79	7.51**
G12	7.86	1.35	0.05	G42	2.76	2.96	1.00**
G13	3.89	1.95	-0.09	G43	2.66	2.45	0.15
G14	3.97	6.89	14.56**	G44	9.88	2.01	1.13**
G15	4.18	3.23	4.61**	G45	5.74	1.49	2.99**
G16	7.73	4.50	2.91**	G46	2.58	3.43	0.24
G17	2.55	0.56	0.10	G47	3.99	-1.97	0.51*
G18	4.81	2.38	-0.12	G48	3.08	-0.14	4.88**
G19	3.55	3.31	1.34**	G49	4.52	2.49	5.44**
G20	7.44	3.42	0.44**	G50	5.59	0.74	12.88**
G21	2.84	-3.27	14.10**	G51	3.33	4.13	62.30**
G22	5.03	2.81	0.09	G52	4.67	-1.75	2.63**
G23	3.77	-2.63	0.59*	G53	3.93	1.59	1.82**
G24	5.16	3.55	-0.09	G54	2.46	2.83	16.89**
G25	2.85	2.65	1.89**	G55	3.38	-0.32	-0.07
G26	8.29	-0.24	7.68**	G56	3.58	0.51	4.72**
G27	7.29	1.00	-0.06	G57	4.41	2.09	1.82**
G28	5.52	5.12	9.99**	G58	5.78	2.55	8.39**
G29	4.54	0.002	9.65**	G59	4.18	2.52	2.66**
G30	7.53	1.66	-0.14	G60	6.26	-2.58	9.45**
				Mean	**5.05**	**SE :**	**0.24**

* Significant at P = 5% ** Significant at P = 1%

(Source: Thirugnanakumar, 1991)

A critical appraisal of the stability and productivity of the individual genotypes on the basis of the three stability parameters has led to the grouping of the sesame genotypes (Thirugnanakumar, 1991) as follows:

The linear as well as non-linear components of G × S interaction for seed yield was non-significant for four genotype *viz.,* G3, G12, G27, G30 (Group I, Table 28). The behaviour of 17 genotypes were not predictable (Group IV). The four genotypes of group I deserve utmost merit consideration for selection for further study to evolve superior stable varieties with high seed yield in sesame.

Table 28 : Grouping of sesame genotypes based on stability parameters for sad yield in sesame

Group I	Group II	Group III	Group III
G3, G12, G27, G30	-	-	G7, G11, G16, G20, G26, G31, G32, G34, G35, G38, G39, G40, G44, G45, G50, G58, G60

(Source: Thirugnanakumar, 1991)

4.5.6. Intercorrelation among stability parameters

Grafius (1956) emphasized that the studies of individual seed yield component could lead to simplification in genetic explanation of yield stability. If traits associated with high yield stability could be found, the plant breeder might effectively select for seed yield stability by selecting for these correlated traits. For that genetic association of the component traits with seed yield should be known.

Seed yield exhibited highly significant positive association with plant height at maturity, number of branches, number of flowers, number of capsules, 1000 seed weight, seed density, total dry matter production and harvest index (Tables 8-11). Thirugnanakumar (1991) studied the intercorrelation among 'b's of seed yield and its component characters as well as among $\bar{s}^2d$'s of seed yield and its component characters (Tables 29 and 30). The 'b' of seed yield was significantly and negatively associated with the 'b' of days to maturity and capsule volume. This indicated that the responsiveness of the genotypes with reference to these characters with the changing seasons had negative impact with the responsiveness of the genotypes for seed yield with the changing seasons. However intercorrelation among the $\bar{s}^2d$ of

Table 29 : Correlation coefficients among 'b's of different characters

Characters	X_2	X_3	X_4	X_5	X_6	X_7	X_8	X_9	X_{10}	X_{11}	X_{12}	X_{13}	X_{14}	X_{15}	X_{16}
X_1	0.3567**	0.2447	0.1893	-0.0112	0.3055*	-0.2358	-0.0019	-0.0777	-0.0378	-0.0534	0.2649*	0.3802**	-0.3983**	0.1184	0.1510
X_2	1.0000	0.1341	-0.1193	-0.1926	0.3872**	-0.5425**	0.0959	0.0197	-0.1373	-0.0567	-0.1228	0.1207	-0.1412	0.2696*	0.6942**
X_3		1.0000	-0.0151	-0.0109	0.0329	-0.0693	0.2820*	0.0484	-0.0919	-0.0408	0.1670	0.1602	-0.1077	0.2403	0.1365
X_4			1.0000	0.1676	-0.0796	0.0730	-0.1500	0.0645	0.1948	-0.0705	0.9777	0.0060	-0.1532	-0.2594*	0.1088
X_5				1.0000	-0.2939*	0.1849	-0.0004	-0.1177	0.1849	-0.0928	-0.1820	-0.3651**	-0.1535	-0.3622**	0.1131
X_6					1.0000	-0.8379**	0.2663*	-0.0052	-0.1153	-0.0823	0.1108	0.2836*	-0.3094*	0.3121*	0.0396
X_7						1.0000	-0.2463	-0.0343	0.0775	0.1064	0.1086	-0.1015	0.2355	-0.3714**	-0.1812
X_8							1.0000	0.1746	-0.0314	0.0114	0.0805	0.1123	-0.0723	0.1524	-0.0218
X_9								1.0000	-0.0139	0.1625	0.1346	0.0631	0.0953	0.1177	-0.0506
X_{10}									1.0000	0.1523	-0.0353	-0.2140	0.2243	-0.3394**	0.0797
X_{11}										1.0000	0.1446	0.0354	-0.0049	-0.1223	0.1166
X_{12}											1.0000	0.7250**	-0.1133	0.0763	-0.0696
X_{13}												1.0000	-0.1041	0.2830	0.0104
X_{14}													1.0000	0.0417	-0.1128
X_{15}														1.0000	-0.1591
X_{16}															1.0000

* Significant at P = 5% level ** Significant at P = 1% level (Source: Thirugnanakumar, 1991)

Table 30 : Correlation coefficients among $\bar{S}^2$ of different characters in sesame

Characters	X_2	X_3	X_4	X_5	X_6	X_7	X_8	X_9	X_{10}	X_{11}	X_{12}	X_{13}	X_{14}	X_{15}	X_{16}
X_1	0.1497	0.1048	-0.0458	-0.1121	0.3063*	0.3498**	-0.2366	0.1311	-0.1193	-0.0141	-0.1201	0.0665	-0.1019	-0.0840	0.0834
X_2	1.0000	-0.0396	0.0247	-0.0253	0.1887	0.7010**	-0.1639	0.0390	-0.0013	-0.0459	-0.0835	0.4529**	-0.0381	0.4043**	0.4611**
X_3		1.0000	0.0214	-0.1523	0.1379	0.0155	-0.0470	-0.0765	-0.3039*	0.0358	0.0261	-0.2655*	0.0117	-0.2736*	-0.0476
X_4			1.0000	-0.0274	0.0132	0.0181	-0.0961	0.2980*	0.0915	-0.0796	0.0217	-0.0146	-0.0483	0.0094	0.0215
X_5				1.0000	0.0729	0.1025	0.0730	0.0064	0.2533*	-0.0624	-0.1434	-0.1113	-0.1139	-0.0251	-0.0881
X_6					1.0000	0.5880**	-0.1435	0.3003*	-0.1739	-0.0391	0.0463	-0.1109	-0.0033	0.0748	0.0563
X_7						1.0000	-0.0941	0.1478	-0.0661	-0.0958	-0.0570	0.2308	-0.0867	0.2275	0.2132
X_8							1.0000	-0.0613	0.0335	-0.0632	0.1326	-0.0343	-0.1353	-0.0802	-0.1752
X_9								1.0000	0.0542	-0.0323	-0.1171	-0.0525	0.0328	0.1727	0.0364
X_{10}									1.0000	0.2555*	-0.0450	0.0918	-0.1133	0.0494	0.2024
X_{11}										1.0000	-0.0307	-0.0817	-0.0872	-0.0675	0.0796
X_{12}											1.0000	-0.1062	-0.1434	0.0994	0.1572
X_{13}												1.0000	0.0206	0.1662	0.2769*
X_{14}													1.0000	0.1617	-0.0600
X_{15}														1.0000	0.3322**
X_{16}															1.0000

* Significant at P = 5% level ** Significant at P = 1% level (Source: Thirugnanakumar, 1991)

seed yield, days to maturity and capsule volume were non-significant, indicating that the stability of seed yield was not affected by the stability of days to maturity and capsule volume. Stability and responsiveness of seed yield were not associated with the stability and responsiveness of the component characters of seed yield. Finlay (1963, 1971), Perkins and Jinks (1968b) and Bucio-Alanis *et al.* (1969) have demonstrated that stability index is a heritable trait in plants. It is well known that yielding ability (mean) and response to environmental changes are two independent attributes of a genotype and are governed by separate sets of gene systems (Singh and Gupta, 1984). Thirugnanakumar (1991) stated that mean, productivity, production response and production stability are not related and relatively independent of each other.

Thirugnankumar (1991) observed that almost all the genotypes had significant 'b's values for all the 16 traits studied. This suggested that the degree of adaptation of different genotypes varied, leading to significant G × S (linear) mean square. Any generalization regarding the stability of a genotype for all the traits is too difficult, since the genotypes studied did not exhibit uniform response and stability patterns. These two attributes namely, responsiveness and stability appeared to be specific for individual character within a given genotype and were not common for all the characters of that genotype. This may be explained on the basis of compromises and compensations among the developmental patterns of the different traits (Singh and Singh, 1980). Literature suggest the importance of component compensation in imparting homeostasis for complex traits like seed yield.

Chapter 5

Genetic Divergence

Genetic architecture of a population is the result of prolonged natural selection. The populations which exist in diverse environments might have been strongly diversified genetically. So, it is necessary to understand the extent and rates of genetic divergence existing between the diversified forms or types. Genetic diversity may be available in improved germplasm or it may exist only in genetically inferior stocks.

Genetic divergence among the parents is important in self pollinated crops because a cross involving genetically diverse parents is likely to produce high heterozygous as well as heterotic effect. In such a cross, more variability could be expected in the segregating generations due to the accumulation of the genes of both the parents. Genetic divergence, as one of the criteria of selection of parents is considered in plant breeding as early as 1962 (Murthy *et al.*, 1962; Timothy, 1963). Later, Joshi and Dhawan (1966), Murthy and Arunachalam (1966), Arunachalam *et al.* (1984) and Lefort-Buson (1987) also stated that for exploiting heterosis as a means of increasing production, it is necessary to have parents of maximum divergence. The more diverse the parents, more is the chance of pronounced heterotic effects and increased spectrum of variability in the segregating generations.

The availability of statistical tools to quantitatively measure the genetic divergence between two or more populations and the relative contribution of individual characters to the total divergence have permitted to trace the evolutionary patterns in some crops such as rice and tobacco and in choosing the parents for hybridization in crop plants (Rao, 1958; Blackith, 1960; Morishima and Oka, 1960; Murthy *et al.*, 1962). Among the several statistical methods developed for measuring the divergence between the populations, multivariate analysis (D^2 statistic) developed by Mahalanobis in 1936, has been found to be a potent tool. Mahalanbobis's D^2 statistic is an effective tool in quantifying the degree of divergence at genetic level and provides a quantitative measure of the association between geographic and genetic diversity based on generalized distance. It is now known that the estimates of genetic divergence are also influenced by environments and seasons. Therefore, if the divergence analysis is conducted in a number of seasons and parents are selected on the basis of consistent divergence over the seasons, the problem of influence of seasons on divergence analysis can be largely overcome.

5.1. Wilk's lambda criterion

The test of significance of difference with regard to the pooled effect of all the characters using Wilk's lambda criterion should be carried out before taking up the study of genetic divergence. This is estimated using the formula given by Wilks (Rao, 1952).

$$\text{Wilk's criterion} = \frac{|E|}{|E+V|}$$

where,

$|E|$ = Determinant of error sum of squares and sum of products matrix

$|E+V|$ = Determinant of (genotype + error) sum of squares and sum of product matrix.

Then the value of 'V' statistic is worked out as follows (Singh and Choudhary, 1985).

'V' stat = -moge

where,

$$m = n - \frac{(p+q+1)}{2}$$

p = number of characters

q = number of genotypes 1

n = degrees of freedom for error + degrees of freedom for genotypes

The value of 'V' statistic is compared with the tabulated chi-square value for pq degrees of freedom and thus significance is tested.

5.2. D^2 analysis

D^2 statistic is used for estimating the genetic divergence among the genotypes, as suggested by Mahalanobis (1936). The data are subjected to multivariate analysis (Rao, 1952; Method-B1). Group constellations of clusters is done as outlined in Singh and Choudhary (1985).

5.3. Intra and inter-cluster distances

After establishing the clusters, the intra-cluster distances are worked out by taking the average of the component genotypes in that cluster. The average inter-cluster divergences are arrived at by taking into consideration, all the component D^2 values possible among the members of the two clusters considered. The square root of the average D^2 values will give the genetic distance 'D' between the clusters. Based on 'D' values (inter-cluster distances) the following scale for rating the distance is adopted (Rao, 1952).

Table 31 : Rating genetic distance

S. No.	Category	'D' values
1.	Less divergent (L)	99 and below
2.	Moderately divergent (M)	Between 100 and 250
3.	Highly divergent (H)	Above 200

(Source: Rao, 1952)

The D^2 values (inter-cluster distances), may also be rated as per the method given by Arunachalam and Bandyopadhyay (1984).

Table 32 : Rating genetic distance

S. No.	Category	Divergent class
1.	DC 1	$D^{2\ 3}$ (m + s)
2.	DC 2	$D^2 <$ (m + s) and $\geq$ m
3.	DC 3	$D^{2\ 3}$ (m-s) and $<$ m
4.	DC 4	$D^2 <$ (m-s)

(Source: Arunachalam and Bandyopadhyay, 1984)
m = mean divergent values s = standard deviation of divergent values

5.4. Example

Thirugnanakumar (1991) has undertaken a study to collect information on genetic diversity in 60 genotypes of sesame originating from different eco-geographic regions of the world, over three different seasons. He has made both individual and pooled analysis. The clustering pattern depicted in Table 33.

Table 33 : Clustering pattern in sesame

Genotypes and Origin	**Seasons**			
	S I	**S II**	**S III**	**Pooled analysis**
Tamil Nadu				
G2	VII	I	I	I
G3	II	I	I	I
G4	IV	I	I	I
G5	II	I	I	I
G6	II	I	I	I
G11	VIII	I	I	I
G12	II	I	I	I
G13	IX	I	I	II
G15	II	I	I	I
G20	I	I	V	II
G27	I	I	I	I
G28	II	I	I	I
G29	I	I	I	I
G30	II	I	I	I
G31	II	I	VI	I
G32	I	I	I	I
G33	II	I	I	I
G34	II	I	I	I
G40	I	I	I	I
G44	II	IV	I	I
G52	I	I	I	I
G53	II	I	I	I
G54	II	I	VIII	I
G55	I	I	I	I
G56	XII	I	II	V
G57	VI	II	II	III

Contd...

G58	I	I	I	I
G59	II	I	I	I
Andhra Pradesh				
G7	II	I	I	I
G8	II	I	I	I
G18	II	I	I	I
G36	II	I	I	I
G42	II	I	I	I
Gujarat				
G1	IV	I	I	I
G26	I	III	I	I
G41	I	II	I	II
G49	II	I	I	I
Maharashtra				
G16	II	I	I	I
G17	V	I	IV	VI
G24	II	I	I	I
Rajasthan				
G48	XI	I	II	IV
Punjab				
G23	X	II	I	III
Haryana				
G46	VI	V	I	II
Ceylon				
G14	V	I	III	II
Pakistan				
G10	II	II	I	I
Japan				
G45	XI	I	VII	IV
Tanzania				
G50	II	I	I	I
Venezuela				
G25	II	I	I	I
Israel				
G37	I	I	I	I

Contd...

Greece				
G35	I	I	I	I
Non-traceable				
G9	II	I	I	I
G19	II	I	I	I
G21	I	I	I	I
G22	I	I	I	V
G38	II	I	I	I
G39	I	I	I	I
G43	III	II	I	I
G47	I	I	I	I
G51	III	VI	I	III
G60	III	I	I	I

(Source: Thirugnanakumar, 1991)

5.4.1. Clustering pattern

D^2 analysis of the 60 genotypes (Table 33) confirmed the presence of high diversity among the genotypes by their resolution into 12 clusters in SI, six clusters in SII and pooled analysis and eight clusters in SIII. Genotypes of different eco-geographic origin were grouped in a single cluster as well as in different clusters.

Cluster I in SI consisted 16 genotypes originating from Tamilnadu (eight genotypes), Gujarat (two genotypes), Israel (one genotype), Greece (one genotype) and four other genotypes, for which the origin is not traceable. In SII, a total of 51 genotypes representing Tamilnadu (26 genotypes), Andhra Pradesh (five genotypes), Gujarat (two genotypes), Maharashtra (three genotypes) and only one genotype from each of Rajasthan, Ceylon, Japan, Tanzania, Venezuela, Israel, Greece and eight genotypes from non-traceable origin were clustered in cluster I. In SIII also, cluster I encompassed 51 genotypes, inclusive of 23 genotypes of Tamilnadu, five genotypes of Andhra Pradesh, four genotypes of Gujarat, two genotypes of Maharashtra and one genotype from each of Punjab, Haryana, Pakistan, Tanzania, Venezuela, Israel, Greece and 10 genotypes of non-traceable origin. But, in the pooled analysis only 47 genotypes were grouped in cluster I. It consisted of 24 genotypes of Tamilnadu, five genotypes of Andhra Pradesh, three genotypes of Gujarat, two genotypes of Maharashtra, one genotype from each of Pakistan, Tanzania, Venezuela, Israel and Greece. It also had eight genotypes of non-traceable origin. The other clusters are also consisted genotypes of varying origins.

The Table 33 also illustrates that the 28 genotypes of Tamilnadu origin were scattered in different clusters in the different seasons and pooled analysis. Those from Tamilnadu might also be from different sources. In SI, out of that 28 genotypes, eight genotypes were found in cluster I, 14 in cluster II, one genotype in each of the clusters IV, VI, VII, VIII, IX, X and XII. In SII, 26 genotypes of Tamilnadu origin were grouped in cluster I, and one in each of clusters II and IV. In SIII, 23 genotypes of Tamilnadu origin were grouped in cluster I, two in cluster II and one in each of the clusters V, VI and VIII. Similarly, in the pooled analysis, 24 genotypes of Tamilnadu origin in the cluster I, two in cluster II and one in each of the clusters III and V were observed. Such a similar pattern of distribution was also found with the genotypes of Gujarat and Maharashtra. This indicated the presence of wide range of genetic diversity among the types of Tamilnadu, Gujarat and Maharashtra. These genotypes can well be exploited in heterosis breeding.

The tendency of the genotypes from diverse geographic regions to group together in one cluster or scattered distribution of genotypes of same origin in different clusters has been inferred from the Table 33. It indicated that the geographical distribution and genetic diversity could not be directly related. Murthy (1965) in tobacco; Murthy and Arunachalam (1966) in linseed, wheat and sorghum; Singh and Gupta (1968) in cotton; Ranga Rao *et al.* (1980) in safflower, John Joel (1987) in sesame, Anandan *et al.* (2011) in rice obtained such a result. Genetic drift and selection in different environments could cause greater diversity than geographical distance. However, geographical isolation could be important in the alteration of the breeding structure as pointed out by Wallace (1963).

In certain cases, the effect of geographic origin appeared to have influenced the clustering pattern. Five genotypes collected from Andhra Pradesh were clustered in a single cluster only, in all the analysis. The only one genotype of Rajasthan (G48) and only one genotype of Japan (G45) were mostly grouped in a single cluster. Such an effect of geographic origin on clustering pattern has been recorded by Jeswani *et al.* (1970) in linseed, Ziauddin Ahamad *et al.* (1980) in triticale and John Joel (1987) in sesame. It suggests that although geographical distribution could not be taken as a sole criterion of genetic diversity, the influence of the former could still be traced.

5.4.2. Intra- and intercluster distances and cluster mean

Cluster III including three genotypes in SI, cluster I consisting 51 genotypes in SII, cluster III representing three genotypes in SIII and cluster V having two genotypes in pooled analysis showed the least intracluster divergence in the concerned analysis (Tables 34-37). This indicated the closeness of the genotypes. This may be explained on the basis that yield being a complex character of polygenic inheritance, similar phenotypes could be produced by many different combinations of genes and such combinations may have similar selective advantage. Further, under constant selection on the segregating population such similar types are expected to be established.

The highest intracluster divergence was recorded for the cluster XI in SI, cluster II in SII, cluster I in SIII and cluster III in the pooled analysis. Limited gene exchange between the genotypes or selection for diverse characters could be responsible for such high intracluster divergence, the performance and expressivity of the genotypes have also to be considered in a given set of season, as the cluster and the composition differed widely season after season.

Highest intercluster distance was observed between the clusters VII and XI in SI. Cluster VII exhibited higher mean values for number of branches (X_3) and seed density (X_{12}) (Table 38). Cluster XI expressed maximum mean value for 1000 seed volume. So, seed yield could well be improved by crossing the genotypes which were grouped in the clusters VII and XI.

In SII, clusters III and VI exhibited highest inter-cluster distance. Cluster III showed maximum mean value for plant height at maturity (X_2), number of branches (X_3), number of flowers (X_4), number of capsules (X_5), oil content (X_{13}) and total dry matter production (X_{15}) (Table 39). Cluster VI expressed minimum value for days to maturity and plant height. So, by effecting crosses among these genotypes, there is a possibility to obtain early maturing but high seed yielding genotypes.

In SIII, cluster VII had maximum mean value for seed breadth (X_9) and 1000 seed volume (X_{11}) (Table 40). Cluster VI had higher mean values for most of the traits which were positively and significantly associated with seed yield. These two clusters also expressed maximum intercluster distance. Hence, there is a possibility to obtain higher seed yielding genotypes by intercrossing among the genotypes of these clusters.

Table 34 : Inter and intra (diagonal) cluster average of D^2 and D (values with in parenthesis) and the extent of diversity among the clusters – S I

Clusters	I	II	III	IV	V	VI	VII	VIII	IX	X	XI	XII
I	**380.6419 (19.51)**	781.4598 (27.95)	867.1466 (29.45)	1404.5502 (37.48)	612.6682 (24.75)	727.0004 (26.96)	1337.2725 (36.57)	942.9822 (30.71)	692.1716 (26.31)	934.8907 (30.57)	3392.0179 (58.24)	1174.3039 (34.27)
II		**343.0669 (18.52)**	673.8246 (25.96)	617.6536 (24.85)	900.2755 (30.01)	1668.7203 (40.85)	767.6436 (27.71)	517.6618 (22.75)	1316.1411 (36.28)	958.3986 (30.96)	5418.5959 (73.61)	2641.6388 (51.40)
III			**329.7550 (18.16)**	1299.661 (36.05)	1440.1532 (37.95)	1246.8694 (35.31)	907.5542 (30.13)	789.5700 (28.10)	1940.910 (44.06)	663.8875 (25.77)	5713.7495 (75.59)	2691.0563 (51.88)
IV				**412.9400 (20.32)**	1474.1323 (38.39)	2378.8845 (48.77)	981.9744 (31.34)	831.0914 (28.83)	1869.9245 (43.24)	1728.1690 (41.57)	6389.0018 (79.93)	3587.937 (59.90)
V					**452.0500 (21.26)**	1547.6095 (39.34)	1533.157 (39.16)	972.8355 (31.39)	688.6914 (26.24)	1492.1115 (33.39)	3651.0005 (60.42)	1655.6265 (40.69)
VI						**448.9100 (21.19)**	2359.5745 (48.58)	1991.3140 (44.62)	1238.3625 (35.19)	1114.7512 (33.39)	2726.9398 (52.22)	863.3516 (29.38)
VII							**0.00 (0.00)**	560.1706 (23.67)	2381.0500 (48.80)	842.3632 (29.02)	6731.2010 (82.04)	3691.6710 (60.76)
VIII								**0.00 (0.00)**	1861.763 (43.15)	1131.22 (33.63)	6171.851 (78.56)	3224.116 (56.78)
IX									**0.00 (0.00)**	1696.2140 (41.19)	2243.6175 (47.37)	716.8431 (26.77)
X										**0.00 (0.00)**	5431.3120 (73.70)	2387.535 (48.86)
XI											**532.090 (23.07)**	858.0136 (29.29)
XII												**0.00 (0.00)**

Figures bold show intra-cluster average of D^2 and D values. (Source: Thirugnanakumar, 1991)

Table 35 : Inter and intra (diagonal) cluster average of D^2 and D (values with in parenthesis) and the extent of diversity among the clusters – S II

Clusters	I	II	III	IV	V	VI
I	**361.7365** **(19.02)**	745.6643 (27.31)	763.0353 (27.62)	287.3988 (16.95)	849.9277 (29.15)	1674.2882 (40.92)
II		**433.1283** **(20.81)**	1592.9838 (39.91)	807.4367 (28.42)	808.2223 (28.43)	980.5618 (31.31)
III			**0.00** **(0.00)**	540.8708 (23.26)	1478.797 (38.46)	2488.392 (49.88)
IV				**0.00** **(0.00)**	854.9784 (29.24)	1906.0630 (43.66)
V					**0.00** **(0.00)**	738.4385 (27.17)
VI						**0.00** **(0.00)**

Figures bold show intra-cluster average of D^2 and D values. (Source: Thirugnanakumar, 1991)

Table 36 : Inter and intra (diagonal) cluster average of D^2 and D (values with in parenthesis) and the extent of diversity among the clusters – S III

Clusters	I	II	III	IV	V	VI	VII	VIII
I	**618.3064** **(24.87)**	1648.0956 (40.60)	1638.8169 (40.48)	2269.3861 (47.64)	1807.3192 (42.51)	1990.2905 (44.61)	2911.5923 (53.96)	956.9554 (30.93)
II		**549.94** **(23.45)**	1827.1253 (42.74)	3628.555 (60.24)	1296.1407 (36.00)	3680.7313 (60.67)	917.441 (30.29)	2545.6483 (50.41)
III			0.00 (0.00)	1172.687 (34.24)	3008.639 (54.85)	3924.147 (62.64)	2938.749 (54.21)	1167.901 (34.17)
IV				0.00 (0.00)	4362.581 (66.05)	3711.188 (60.92)	4555.343 (67.49)	1773.729 (4212)
V					0.00 (0.00)	1434.842 (37.87)	1935.241 (43.99)	3273.55 (57.21)
VI						0.00 (0.00)	4657.567 (68.25)	2520.048 (50.20)
VII							0.00 (0.00)	3839.323 (61.96)
VIII								0.00 (0.00)

Figures bold show intra-cluster average of D^2 and D values. (Source: Thirugnanakumar, 1991)

Table 37 : Inter and intra (diagonal) cluster average of D^2 and D (values with in parenthesis) and the extent of diversity among the clusters – Pooled

Clusters	I	II	III	IV	V	VI
I	**153.0453 (12.37)**	359.0404 (18.95)	527.9698 (22.98)	1436.7764 (37.90)	516.9902 (22.74)	752.7905 (27.44)
II		**194.6906 (13.95)**	498.8438 (22.33)	732.0309 (27.06)	190.7513 (13.81)	658.4696 (25.66)
III			**549.7900 (23.45**	1496.2470 (38.68)	582.9825 (24.15)	1264.645 (35.56)
IV				**104.0300 (10.20)**	384.5471 (19.61)	1477.9685 (38.44)
V					**88.1400 (9.39)**	681.5823 (26.11)
VI						0.00 (0.00)

Figures bold show intra-cluster average of D^2 and D values. (Source: Thirugnanakumar, 1991)

Table 38 : Cluster means for sixteen characters in sesame – SI

Characters	I	II	III	IV	V	VI	VII	VIII	IX	X	XI	XII
X_1	107.09	106.32	103.67	106.28	<u>115.25</u>	**102.10**	107.20	106.70	106.10	105.40	106.40	107.20
X_2	79.78	80.02	**40.81**	92.30	<u>99.68</u>	75.90	78.08	94.04	83.25	46.09	80.44	85.01
X_3	8.69	9.22	**5.97**	13.75	10.68	9.95	<u>20.80</u>	9.45	8.00	9.65	8.93	6.40
X_4	74.93	79.75	25.43	105.28	88.43	52.33	67.10	<u>112.90</u>	38.00	**22.60**	68.93	55.35
X_5	63.79	68.79	**17.43**	77.00	77.45	42.65	49.40	<u>109.60</u>	29.50	18.40	59.65	45.90
X_6	0.33	0.32	0.30	0.42	<u>0.48</u>	0.30	**0.13**	0.28	0.38	0.22	0.43	0.40
X_7	54.65	54.81	57.82	57.52	58.28	53.56	**50.28**	52.40	**59.12**	53.80	57.65	55.40
X_8	3.03	3.15	3.11	3.36	3.20	**2.89**	3.00	2.94	**3.52**	3.12	3.22	3.11
X_9	1.77	1.82	1.74	<u>1.94</u>	1.86	1.79	1.76	1.77	1.87	**1.63**	1.82	1.70
X_{10}	2.77	2.97	**2.30**	3.36	3.34	2.46	3.49	<u>3.59</u>	3.32	3.10	2.45	2.46
X_{11}	8.28	5.49	5.19	4.98	8.30	9.32	**4.97**	5.93	9.84	5.97	**13.47**	11.05
X_{12}	0.34	0.55	0.46	0.68	0.41	0.27	<u>0.70</u>	0.61	0.34	0.52	**0.19**	0.22
X_{13}	43.09	44.48	42.72	44.00	43.45	37.25	38.70	<u>45.70</u>	45.40	**32.75**	45.15	41.20
X_{14}	4.13	4.60	3.20	4.08	4.44	**2.55**	4.26	<u>9.31</u>	3.04	4.21	4.14	2.85
X_{15}	21.18	20.72	**10.72**	23.60	<u>29.77</u>	17.42	18.35	26.09	13.59	14.04	23.54	18.12
X_{16}	19.22		28.15	17.38	**11.44**	16.02	22.76	<u>35.58</u>	22.47	29.91	27.26	15.83

Bold indicates minimum cluster values. (Source: Thirugnanakumar, 1991)
Underline indicates maximum cluster values.

Table 39 : Cluster means for sixteen characters in sesame – S II

Characters	Clusters					
	I	II	III	IV	V	VI
X_1	105.11	104.12	104.30	105.30	97.50	**94.10**
X_2	53.97	37.94	64.10	56.00	46.61	**32.66**
X_3	8.45	8.14	10.80	9.20	**6.80**	7.00
X_4	76.17	**56.86**	132.00	91.00	61.40	66.70
X_5	57.95	38.60	70.30	64.10	**22.90**	27.75
X_6	0.37	**0.31**	0.35	0.41	0.53	0.39
X_7	55.50	**53.34**	55.80	55.80	56.20	55.60
X_8	3.10	3.00	**2.81**	3.41	3.13	2.83
X_9	1.63	1.63	**1.53**	1.79	1.73	1.56
X_{10}	3.10	3.00	3.28	3.56	3.12	**2.46**
X_{11}	4.77	4.88	4.62	4.72	5.31	**3.65**
X_{12}	0.65	0.62	0.71	0.76	**0.59**	0.68
X_{13}	42.95	37.74	45.75	43.65	41.15	**31.15**
X_{14}	5.12	3.64	6.80	7.29	**2.42**	3.65
X_{15}	21.64	20.53	33.50	23.19	**17.11**	21.99
X_{16}	24.78	19.03	20.44	31.52	**14.23**	16.42

Bold indicates minimum cluster values. Underline indicates maximum cluster values.
(Source: Thirugnanakumar, 1991)

Table 40 : Cluster means for sixteen characters in sesame – S III

Characters	Clusters							
	I	II	III	IV	V	VI	VII	VIII
X_1	96.95	93.77	116.45	<u>129.30</u>	92.90	99.40	**92.00**	100.00
X_2	50.83	41.89	46.85	**35.70**	64.02	<u>72.90</u>	43.83	37.71
X_3	8.80	8.63	7.75	**7.30**	<u>11.00</u>	10.90	8.35	9.30
X_4	76.97	65.33	**54.70**	60.50	117.20	<u>144.30</u>	79.10	64.20
X_5	53.77	37.13	17.10	**11.90**	102.65	<u>124.55</u>	18.65	37.15
X_6	0.34	**0.20**	<u>0.39</u>	0.29	0.36	0.29	0.22	0.29
X_7	50.18	**48.47**	51.00	50.60	51.00	49.20	48.90	<u>51.60</u>
X_8	<u>3.03</u>	2.81	2.78	2.85	2.85	<u>3.02</u>	2.95	**2.68**
X_9	1.65	1.60	1.68	**1.48**	1.75	1.54	<u>1.81</u>	1.51
X_{10}	3.81	4.72	**2.18**	2.79	<u>5.39</u>	4.65	3.22	3.42
X_{11}	6.66	11.54	8.32	8.90	10.46	5.06	<u>14.57</u>	**3.79**
X_{12}	0.60	0.43	0.27	0.32	0.51	<u>0.92</u>	**0.22**	0.90
X_{13}	42.22	39.85	**32.70**	42.30	41.70	<u>44.90</u>	43.90	38.35
X_{14}	5.84	4.29	0.68	**0.62**	13.82	<u>17.52</u>	1.79	0.99
X_{15}	21.09	20.21	12.54	**7.48**	40.81	<u>41.33</u>	9.97	7.83
X_{16}	26.22	20.87	**5.59**	8.30	33.83	<u>42.51</u>	17.94	12.78

Bold indicates minimum cluster values. Underline indicates maximum cluster values.
(Source: Thirugnanakumar, 1991)

In the pooled analysis, cluster III evinced minimum number of days to maturity and cluster IV had maximum mean value for capsule volume (X_6), seed length (X_8), seed breadth (X_9), 1000 seed volume (XII) and oil content (X_{13}) (Table 41). These two clusters also showed maximum inter-cluster distance. By inter-crossing among the genotypes of these clusters, there is possibility to obtain high yielding genotypes.

Table 41 : Cluster mean for 16 characters in sesame – pooled

S. No.	Characters	Clusters					
		I	II	III	IV	V	VI
1	X_1	102.79	103.38	**99.57**	100.35	103.85	119.70
2	X_2	61.66	59.11	**40.28**	54.43	61.38	58.63
3	X_3	8.95	8.12	**7.67**	8.25	8.85	10.60
4	X_4	78.30	73.80	**49.40**	68.80	69.60	64.90
5	X_5	60.90	47.34	42.90	43.70	48.75	**33.70**
6	X_6	0.32	0.38	0.28	0.74	**0.27**	0.37
7	X_7	54.04	54.20	**52.40**	54.05	53.50	54.30
8	X_8	3.08	3.06	**2.90**	3.13	3.07	3.01
9	X_9	1.70	1.73	**1.60**	1.75	1.66	1.64
10	X_{10}	3.31	3.17	3.08	**3.06**	3.24	3.15
11	X_{11}	**5.77**	7.74	5.83	10.88	8.80	7.60
12	X_{12}	0.60	0.45	0.56	**0.35**	0.41	0.45
13	X_{13}	43.12	41.14	**36.60**	44.15	41.20	42.90
14	X_{14}	5.32	3.93	3.84	4.41	4.31	**2.55**
15	X_{15}	21.31	21.82	18.60	21.46	18.59	**16.25**
16	X_{16}	24.58	17.32	21.64	22.30	26.48	**16.24**

Bold indicates minimum cluster values.
Underline indicates maximum cluster values. (Source: Thirugnanakumar, 1991)

As pointed out by Sokal (1965), the choice of the characters is important in multivariate analysis. Cyclic inbreeding and crossing may be helpful in bringing new genes into a population of any crop and thus, expanding the range of adaptation. As brought out by Mather (1943), polygenic variability which is necessary for prospective, adaptive and evolutionary change need not exist as true phenotypic variation which effect fitness. Therefore, phenotypic uniformity with genetic diversity within the population appears to very useful in any crop. In fact, this is possible with the proper use of the above suggested parents in the formation of the gene complexes to replace the existing varieties.

From the foregoing discussion, it may be stated that there is no obvious relationship between genetic and geographical diversity. Inter and intra-cluster distances were also not related. The grouping pattern was not consistent across seasons, indicating that genetic diversity do depend on the season and that genotype ´ season interaction affect genetic diversity. It is highly desirable that the extent of genetic divergence between the populations reflected through any analysis, should be comparatively stable over seasons and environments, to be of utility to plant breeders. The differential grouping of different genotypes of varying origin in different seasons, does not mean that a breeder should study divergence in only one season. In fact, it really suggests the need for studying the divergence in varying seasons.

5.4.3. Selection of parents

Singh and Gupta (1984); Singh *et al.* (1988) and Thirugnanakumar (1991) suggested three approaches to select the parents based on the genetic divergence in varying seasons.

(a) One may select the parents on the basis of divergence exhibited in the richest and most productive season. Because, it provides opportunity for the fullest expression of the genetic potential of a genotype. The relative ranking of the three seasons, revealed that SIII was best for days to maturity (X_1), number of flowers (X_4), 1000 seed weight (X_{10}), 1000 seed volume (X_{11}), seed yield (X_{14}) and harvest index. Using this criterion, the divergent genotypes in SIII may be selected from different clusters having high inter-cluster distances. Clusters VI and VII showed maximum inter-cluster distance in SIII. Only one genotype in each of these two clusters *viz.*, G31and G45 were found, respectively.

(b) Selection of parents can also be made on the basis of divergence which is consistent over the seasons. This can be taken as a reliable indication of the genetic divergence. The following genotypes showed consistent divergence in all the three seasons.

(i) G56 was consistently divergent with the following genotypes: G10, G23, G26, G41, G44 and G46.

(ii) G45 and G48 were consistently divergent from G10, G23, G26, G41, G44 and G46.

(iii) G14 and G17 were consistently divergent from G10, G23, G26, G41, G44, G46 and G57.

(iv) G23 was consistently divergent from G14, G17, G31, G45 and G54.

(v) G46 was consistently divergent from G14, G17, G20, G31, G45, G48, G54 and G56.

The divergent genotypes *viz.,* G14, G17, G45, G48 and G56 were also consistently divergent from the divergent genotypes G23 and G46. But, the divergence among these two groups was not consistent.

(c) One may also argue that the divergence expressed in the pooled analysis may be a reliable estimate and therefore should be used for selecting the parents. Accordingly, the genotypes *viz.,* G13, G14, G20, G41 and G46 were divergent from the rest of the sixty genotypes investigated. G23 and G51 were divergent from all other fifty eight genotypes. The genotypes *viz.,* G45 and G48 were divergent from the remaining fifty eight genotypes. Similarly, the genotypes G22 and G56 were divergent from the remaining fifty eight genotypes.

5.4.4. Character contribution towards genetic diversity

Relative ranking of the contribution of the 16 characters towards genetic diversity based D^2 statistic revealed a high genotype × season interaction in their percentage contribution towards genetic diversity. Volume of 1000 seeds followed by oil content exhibited highest contribution towards genetic diversity in SI, SIII and pooled analysis. In SII, oil content followed by number of capsules exhibited the highest contribution. The highest contributor namely, volume of 1000 seeds scored eleventh rank in SII (Table 42). Even though oil content scored second rank in three seasons, its contribution was constant across seasons and pooled over seasons.

Table 42 : Relative ranking of characters contributing towards diversity (Based on D^2 statistic)

Rank	SI	SII	SIII	Pooled
1.	Volume of 1000 seeds	Oil content	Volume of 1000 seeds	Volume of 1000 seeds
2.	Oil content	Number of capsules	Oil content	Oil content
3.	Length of seeds	Length of seeds	Number of capsules	Days to maturity
4.	Density of seeds	Volume of capsules	Days to maturity	Number of flowers
5.	Days to maturity	Number of flowers	Seed yield	Volume of capsules
6.	Weight of 1000 seeds	Days to maturity	Number of flowers	Length of seeds
7.	Number of flowers	Breadth of seeds	Total dry matter production	Density of seeds
8.	Volume of capsules	Total dry matter production	Harvest index	Number of capsules
9.	Number of branches	Weight of 1000 seeds	Volume of capsules	Seed yield
10.	Number of capsules	Number of seeds per capsule	Length of seeds	Weight of 1000 seeds
11.	Breadth of seeds	Volume of 1000 seeds	Density of seeds	Total dry matter production
12.	Total dry matter production	Height of the plant at maturity harvest index	Number of seeds per capsule	Number of branches
13.	Harvest index	Number of branches	Height of the plant at maturity	Number of seeds per capsule; harvest index; Height of the plant at maturity; Breadth of seeds
14.	Number of seeds per capsule; seed yield	Density of seeds; seed yield	Weight of 1000 seeds	
15.	Height of the plant at maturity		Breadth of seeds; number of branches	

Chapter 6

Combining Ability

6.1. Line × tester analysis

A successful approach of crop breeding programme depends on the proper choice of best parents for hybridization and the ideal selection adopted in the early generation. But, the selection of parents for hybridization programme is relatively tough in the case of complex traits like yield and their components as they governed by large number of quantitative genes which are influenced by environments and seasons. Thus, breeders often meet with difficulty to fix in advance the desirable parents which will give superior progenies. In such a situation, knowledge on the nature of gene action on such complex quantitative traits of economic importance is necessary to plan and adopt appropriate selection techniques (Simmonds, 1979) and breeding methodology.

Combining ability analysis gives useful information regarding the selection of parents in terms of the performance of the hybrids. This analysis elucidates the nature and magnitude of various types of gene action involved in the expression of quantitative traits (Dhillon, 1975). Thus, the knowledge of combining ability serves as an useful tool for the selection of plants for hybridization and further exploitation. Combining ability consists of general and specific combining ability.

Sprague and Tatum (1942) defined general combining ability (*gca*) as the "average performance of a line in hybrid combinations" and specific combining ability (*sca*) "as those cases in which certain combinations do relatively better of worse than would be expected on the basis of the average performance of the lines involved. The general combining ability provides an indication of the importance of genes of largely additive in nature, while specific combining ability indicates the importance of non-additive gene effects.

Several biometrical techniques are available to estimate the general combining ability of parents and specific combining ability of hybrids. Among them, line × tester analysis is extensively used both in self and cross pollinated crops. The advantage of line × tester analysis is that a large number of inbred lines can be evaluated than diallel analysis. Besides, it provides additional information on the performance of the specific hybrids on gene action in contrast to top cross or polycross. The line × tester analysis is an efficient biometrical approach for assessing the combining ability of parents and hybrids (Kempthorne, 1957).

6.1.1. Estimation of combining ability

The analysis of combining ability is carried out following Kempthorne's method (1957) is follows. The general combining ability of the parents and specific combining ability of the hybrids are assessed. The mean squares due to different sources of variation in each experiment as well as their genetic expectations are estimated as indicated in the ANOVA table given below.

Source	**Degrees of freedom**	**Mean squares**	**Expectations of squares**
Replications	(r – 1)		
Hybrids	(lt – 1)		
Lines	(l – 1)	M_1	s^2e + r [Cov. (F.S.) – 2 Cov.(H.S)] + rt [Cov.(H.S)]
Testers	(t – 1)	M_2	s^2e + r [Cov. (F.S.) – 2 Cov.(H.S)] + rl [Cov.(H.S)]
Lines × testers	(l – 1) (t – 1)	M_3	s^2e + r [Cov. (F.S.) – 2 Cov.(H.S)]
Error	(r – 1) (lt – 1)	M_4	s^2e
Total	**(ltr – 1)**		

(Source: Kempthorne, 1957)

r = Number of replications l = Number of lines t = Number of testers

Estimation of *gca* and *sca* effects

The *gca* and *sca* effects in each environment were estimated. The analysis was done on the following model:

$$X_{ijk} = \mu + g_i + g_j + s_{ij} + e_{ijk} + r_k$$

where,

X_{ijk} = Value of the ijkth observation

m = Population mean

g_i = gca effect of the ith line

g_j = gca effect of the jth tester

s_{ij} = sca effect of the ijth hybrid

e_{ijk} = error effect associated with ijkth observation

i = number of lines

j = number of testers

k = number of replications

r_k = kth replication effect

The individual effects of gca and sca were obtained from the two way table of lines Vs. testers, in which each figure was a total over replications, as follows:

$$m = \frac{X...}{rlt}$$

$$\hat{g}_i = \frac{X_{i..}}{rt} - \frac{X...}{rlt}$$

$$\hat{g}_i = \frac{X_{.j.}}{rl} - \frac{X...}{rlt}$$

$$\hat{s}_{ij} = \frac{X_{ij.}}{r} - \frac{X_{i}..}{rt} - \frac{X_{.j.}}{rl} - \frac{X...}{rlt}$$

Cov (F.S.) = Covariance between full sibs

Cov (H.S.) = Covariance between half sibs

Estimates of covariance of full sibs and that of half sibs were calculated from the genetic expectations of mean squares as:

$$\text{Cov. (F.S.)} = \frac{M_1 + M_2 + M_3 - 3M_4 + 6r\ \text{Cov.(H.S.)} - r(1+t)\text{Cov.(H.S.)}}{3r}$$

$$\text{Cov. (H.S.)} = \frac{M_1 + M_2 - 2M_3}{r(1+t)}$$

From the covariance of full sibs and covariance of half sibs, variances due to general combining ability (s^2gca) and specific combining ability (s^2sca) were estimated as follows:

$\sigma^2 gca = Cov.H.S.$

$\sigma^2 sca = Cov.F.S. - 2\ Cov.H.S.$

Proportional contribution by lines, testers and their interaction to total variance was calculated as per Singh and Chaudhary (1985).

$$\text{Contribution of lines} = \frac{SS(l) \times 100}{SS\ (crosses)}$$

$$\text{Contribution of testers} = \frac{SS(t) \times 100}{SS\ (crosses)}$$

$$\text{Contribution of lines} \times \text{testers} = \frac{SS(l \times t) \times 100}{SS\ (crosses)}$$

where,

$X...$ = total of all hybrid combinations

$X_i..$ = total of ith line over 't' testers and 'r' replications

$X._j.$ = total of jth tester over 'l' lines and 'r' replications

$X_{ij}.$ = total of the hybrids of ith line and j^{th} tester 'r' replications.

The standard errors pertaining to gca and sca effects were calculated from the square root of variance effects as indicated below.

(i) Standard error effects for lines

$$SE(g_i) = \left(\frac{\sigma^2 e}{rt}\right)^{1/2}$$

(ii) Standard error effects for testers

$$SE(g_j) = \left(\frac{\sigma^2 e}{rl}\right)^{1/2}$$

(iii) Standard error effects for hybrids

$$SE(s_{ij}) = \left(\frac{\sigma^2 e}{r}\right)^{1/2}$$

Pooled analysis

To understand the real picture of genetic architecture of the hybrids and their parents, the data of both seasons were subjected to pooled analysis (Vaidyanathan, 1982).

The model used for the analysis was:

$$X_{ijkr} = \mu + l_i + t_j(lt)_{ij} + n_k + (ln)_{ik} + (tn)_{jk} + (ltn)_{ijk} + e_{ijkr}$$

where,

X_{ijkr} = Value of the ijkrth observation

μ = Overall mean

l_j = gca effect of the ith line parent

t_j = gca effect of the jth tester parent

$(lt)_{tj}$ = sca effect of the hybrids

n_k = Effect of the kth season

$(ln)_{ik}$ = ith line and kth season interaction effect

$(tn)_{jk}$ = jth tester and kth season interaction effect

$(ltn)_{ijk}$ = hybrids × season interaction effect

e_{ijkr} = random error effects associated with ijkrth observation

l = number of lines

t = number of testers

n = number of seasons

r = number of replications

The form of pooled analysis of variance is given below.

Source	Degrees of freedom	Mean squares	Expectations of squares
Seasons	(n – 1)		
Replication in seasons	n(r – 1)		
Hybrids	(lt – 1)		
Lines	(l – 1)	M_1	$\sigma^2e + r\sigma^2ltn + rn\sigma^2lt + rtn\sigma^2l$
Testers	(t – 1)	M_2	$\sigma^2e + r\sigma^2ltn + rl\sigma^2tn + rn\sigma^2lt + rln\sigma^2t$
Lines × Testers	(l – 1) (t – 1)	M_3	$\sigma^2e + r\sigma^2ltn + rn\sigma^2lt$

Lines × Seasons	(l – 1) (n – 1)	M_4	$\sigma^2 e + r\sigma^2 ltn + rt\sigma^2 ln$
Testers × Seasons	(t – 1) (n – 1)	M_5	$\sigma^2 e + r\sigma^2 ltn + rl\sigma^2 tn$
Lines × Testers × Seasons	(l – 1) (t – 1) (n – 1)	M_6	$\sigma^2 e + r\sigma^2 ltn$
Error	n(lt – 1) (r – 1)	M_7	$\sigma^2 e$

(Source: Vaidyanathan, 1982)

where,

n = number of seasons

σ^2_l = variance due to lines

σ^2_t = variance due to testers

σ^2_{lt} = variance due to lines × testers interaction

σ^2_{ln} = variance due to lines × season interaction

σ^2_{tn} = variance due to testers × season interaction

σ^2_{ltn} = variance due to (lines × testers) season interaction

Estimation of variance components was done as follows by equating mean squares to expectations and sorting out for the appropriate components.

$\sigma^2_l = (M_1 - M_3 - M_4 + M_6)/rtn$

$\sigma^2_t = (M_2 - M_3 - M_5 + M_6)/rln$

$\sigma^2_{lt} = (M_3 - M_6)/rn$

$\sigma^2_{ln} = (M_4 - M_6)/rt$

$\sigma^2_{tn} = (M_5 - M_6)/rl$

$\sigma^2_{ltn} = (M_6 - M_7)/r$

Likewise, variance components for gca, sca and their respective interaction with the season was calculated as follows:

$$\sigma^2 \text{ gca} = \frac{M_1 + M_2 - 2M_3 - M_4 - M_5 + 2M_6}{rn(l + t)}$$

σ^2 sca $= M_3 - M_6/rn$

σ^2gca × n $= M_4 + M_5 - 2\ M_6/r(l+t)$

σ^2sca × n $= M_6 - M_7/r$

Estimation of general and specific combining ability effects of ijkth observation in case of combined estimate was calculated as in single season, except that the total of all replications over the two conditions was used for the two way table of lines vs. testers and the denominator in all cases was rltn instead of rlt.

For calculation of standard errors effects, the same formulae were used as in case of single season except the denominator which increases by multiplication of number of seasons.

Heterosis

Heterosis values were worked out by utilizing the overall mean of each metric trait. Relative heterosis was estimated as the per cent deviation of the F_1 cross from its mid-parental value. Heterobeltiosis for each character, in each hybrid combination was expressed as the per cent increase or decrease of F_1 value over the standard parent value, for all the characters. Estimation proceeds as follows:

$$d_i = \text{Heterosis \% (over mid-parent)} = \frac{\overline{F}_1 - \overline{MP}}{\overline{MP}} \times 100 \text{ s}$$

6.1.2. Example

Thirugnanakumar (1991) studied the combining ability in sesame using 27 lines × 3 testers. He estimated the *gca* and *sca* effects for 16 traits over two seasons *viz.*, S_1 and S_2. He also computed pooled analysis.

6.1.2.1. Analysis of variance

The analysis of variance within each season (individual season) revealed the presence of significant variation among the hybrids, lines and line × tester interaction for 11 out of 16 characters analyzed (Table 43). The characters *viz.*, capsule volume (X_6), seed length (X_8), seed breadth (X_9), 1000 seed weight (X_{10}) and seed density (X_{12}) exhibited non-significant variances. Additionally, the variance due to line × tester interaction was non-significant for number of braches in S_2. The variance due to testers was also significant for 11 out of 16 traits studied. Capsule volume, seed length, seed breadth and seed density displayed non-significant variances both in S_1 and S_2. Apart from these traits, 1000 seed volume in S_1 and oil content in S_2 showed non-significant variation. The pooled analysis of variance over seasons revealed the presence of significant variation among hybrids, lines, line × tester interaction, line × season interaction and (line × tester) × season interaction for 11 out of 16 traits studied (Table 44). Capsule volume, seed length, seed breadth, 1000 seed weight and seed density exhibited non-significant variances. The variance due to testers, season and testers × season interaction was significant for 12 out of 16 characters studied. Capsule volume, seed length, seed breadth and seed density displayed non-significant variances. This indicated that lines, testers and hybrids were influenced by the seasons. The non-significant variance may atleast partially due to the narrow genetic bases.

6.1.2.2. Combining ability variance

The estimates of combining ability variances showed higher values for SCA variances than GCA variances for all the traits studied in individual analyses (Table 45). This indicated the predominance of non-additive gene action in the inheritance of the 16 traits of interest. However, the variance due to GCA was higher than the SCA variance for seed density in the pooled analysis. This may indicate the existence of both additive and non-additive gene action in the expression of seed density. For all the other traits in the pooled analysis, the variances were not comparable due to negative sign variances of general combining ability and specific combining ability and their interaction with different environments is useful in formulating an efficient breeding programme (Matzinger *et al.,* 1959). The SCA × season interaction variances were higher than GCA × season interaction variances. This indicated that hybrids interacted more than the parents with the seasons.

6.1.2.3. Proportional contribution of lines, testers and line × testers interaction to the total variance

Considering the proportional contribution by lines, testers and line × tester interaction to the total variance, the line × testers interaction contributions were higher than either lines or testers for almost all the characters studied. This clearly indicated that importance of hybrid vigour and non-additive gene action (Table 46).

Table 43 : Analysis of variance for combining ability in individual seasons

Characters	Hybrids (DF 80)		Lines (DF 26)		Testers (DF 2)		Lines × Testers (DF 52)		Error (DF 110)	
	S1	S2	S1	S2	S1	S2	S1	S2	S1	S2
X1	91.39**	42.72	86.83**	57.44**	644.75**	130.69**	72.38**	31.97**	1.36	3.11
X2	486.60**	148.60**	423.06**	297.61**	5885.94**	151.13**	310.70**	74.01**	21.89	27.51
X3	38.93**	2.42**	50.15**	4.18**	236.45**	5.67**	25.72**	1.42	1.10	0.68
X4	627.74**	298.97**	424.85**	549.71**	7506.14**	209.78**	464.63**	177.03**	29.85	31.10
X5	714.74**	414.72**	546.16**	685.73**	10839.55**	853.11**	409.62**	262.35**	25.14	24.31
X6	0.02	0.03	0.01	0.04	0.14	0.12	0.02	0.02	0.001	0.001
X7	4.71**	8.56**	4.09**	8.24**	52.09**	33.08**	3.19**	7.78**	1.28	2.49
X8	0.08	0.08	0.05	0.10	0.50	0.02	0.07	0.07	0.01	0.01
X9	0.04	0.03	0.02	0.04	0.49	0.05	0.03	0.02	0.01	0.01
X10	1.44	0.63	1.23	0.45	10.13**	3.20*	1.22	0.62	0.03	0.01
X11	7.94**	3.40**	7.87**	2.42**	1.73	23.39**	8.22**	3.11**	0.03	0.08
X12	0.08	0.04	0.07	0.05	0.21	0.20	0.08	0.04	0.002	0.003
X13	24.04**	12.35**	22.62**	14.26**	69.42**	0.50	23.00**	11.86**	0.34	1.39
X14	5.13**	4.95**	3.66**	7.25**	41.00**	15.02**	4.49**	3.40**	0.13	0.68
X15	41.83**	44.29**	48.24**	56.18**	209.32**	76.62**	32.18**	37.10**	4.09	4.71
X16	117.45**	101.19**	122.07**	118.13**	501.73**	142.77**	100.36**	91.12**	6.53	19.44

* Significant at P = 5% ** Significant at P = 1%
(Source: Thirugnanakumar, 1991)

Table 44 : Pooled analysis of variance for combining ability

Characters	Seasons (DF 1)	Hybrids (DF 80)	Lines (DF 26)	Testers (DF 2)	Lines × Testers (DF 52)	Lines × Season (DF 26)	Testers × Season (DF 52)	(Lines × Testers) × Season (DF 52)	Error (DF 220)
X_1	55486.25**	57.04**	78.30**	105.88**	44.53**	65.98**	669.56**	59.82**	2.24
X_2	4884.63**	333.87**	371.32**	2077.44**	248.08**	349.35**	3959.63**	136.62**	24.70
X_3	7671.78**	20.75**	29.35**	104.09**	13.24**	24.98**	138.03**	13.91**	0.89
X_4	8183.38**	496.94**	641.55**	2869.97**	333.36**	333.01**	4845.95**	308.30**	30.48
X_5	10606.69**	514.69**	798.29**	3069.69**	274.62**	433.61**	8622.97**	397.35**	24.72
X_6	0.53	0.02	0.03	0.02	0.02	0.02	0.24	0.02	0.001
X_7	423.13**	7.60**	9.03**	25.78**	6.19**	3.30**	59.39**	4.79**	1.88
X_8	0.00	0.08	0.08	0.16	0.08	0.07	0.32	0.07	0.01
X_9	0.89	0.03	0.02	0.24	0.03	0.03	0.31	0.02	0.01
X_{10}	22.71**	0.80	0.50	6.28**	0.73	1.18	7.05**	1.11	0.02
X_{11}	5.23*	6.37**	6.92**	6.26**	6.10**	3.37**	18.86**	5.23**	0.06
X_{12}	1.00	0.07	0.08	0.40	0.06	0.05	0.02	0.05	0.002
X_{13}	3453.19**	20.58**	19.64**	29.75**	20.70**	17.24**	40.17**	14.16**	0.87
X_{14}	188.47**	4.83**	5.42**	3.55*	4.58**	5.49**	52.47**	3.31**	0.41
X_{15}	396.10**	41.47**	41.09**	64.03**	40.80**	63.34**	221.90**	28.48**	4.40
X_{16}	23688.1**	112.54**	109.72**	315.76**	106.13**	130.48**	328.74**	85.35**	12.98

* Significant at P = 5% ** Significant at P = 1%

(Source: Thirugnanakumar, 1991)

Table 45 : Estimates of combining ability variances in individual and pooled analysis and their interaction with the seasons

Characters	Individual						Pooled				
	σ^2 gca		σ^2 sca		σ^2 gca/σ^2 sca		σ^2 gca	σ^2 sca	σ^2 gca × season	σ^2 sca × season	σ^2 gca/ σ^2 sca
	S_1	S_2	S_1	S_2	S_1	S_2					
X_1	9.7803	2.0698	35.51	14.43	0.2754:1	0.1434:1	-4.5186	1.5400	9.9075	28.7900	NE
X_2	94.7933	5.012	144.405	23.25	0.6564:1	0.2156:1	-18.2034	53.1825	66.5745	55.9600	NE
X_3	3.9193	0.1168	12.31	0.3700	0.3184:1	0.3158:1	-0.3331	2.7675	2.0575	6.5100	NE
X_4	116.6955	6.7572	217.39	72.965	0.5368:1	0.0926:1	-14.3101	6.1775	76.0452	138.9100	NE
X_5	176.1078	16.9023	192.24	119.02	0.9161:1	0.1420:1	-42.5178	9.0650	135.0482	186.3150	NE
X_6	0.0018	0.0020	0.0095	0.0095	0.1930:1	0.2105:1	-0.0018	0.00	0.0034	0.0095	NE
X_7	0.8300	0.4293	0.9550	2.6450	0.8691:1	0.1623:1	-0.2316	-0.3725	0.9333	1.4550	NE
X_8	0.0068	-0.0003	0.0300	0.03	0.2278:1	NE	-0.0013	0.00	0.0043	0.0300	NE
X_9	0.0075	0.0008	0.0100	0.0050	0.7500:1	0.1667:1	-0.0008	0.0025	0.0050	0.0050	NE
X_{10}	0.1487	0.0402	0.5950	0.3050	0.2499:1	0.1317:1	-0.0095	0.0175	0.0927	0.5450	NE
X_{11}	-0.114	0.3265	4.095	1.5150	NE	0.2155:1	-0.0672	-0.4650	0.2417	2.585	NE
X_{12}	0.0020	0.0028	0.0390	0.0185	0.0513:1	0.1532:1	0.0033	0.00	-0.0003	0.0240	NE
X_{13}	0.7673	-0.1493	11.3300	5.235	0.0677:1	NE	-0.1470	0.7700	0.5425	6.6450	NE
X_{14}	0.5947	0.2778	2.1800	1.36	0.2728:1	0.1896:1	-0.4370	0.5450	0.8405	1.4500	NE
X_{15}	3.2200	0.9767	14.0450	16.1950	0.2293:1	0.0603:1	-1.8942	8.7150	3.4290	12.0400	NE
X_{16}	7.0513	1.3110	46.9150	35.8400	0.1503:1	0.0366:1	-0.8304	11.2825	4.4028	36.1850	NE

(Source: Thirugnanakumar, 1991)

Table 46 : Proportional percentage contribution by lines, testers and line ´ tester interaction to the total variance

Characters	Lines			Testers			Lines × Testers		
	S_1	S_2	P	S_1	S_2	P	S_1	S_2	P
X_1	30.88	43.70	44.61	17.64	7.65	4.64	51.48	48.65	50.75
X_2	28.26	65.09	36.15	30.24	2.54	15.55	41.50	32.37	48.30
X_3	41.87	56.04	45.98	15.18	5.84	12.54	42.95	38.12	41.48
X_4	22.00	59.76	41.96	29.89	1.75	14.44	48.11	38.49	43.60
X_5	24.84	53.74	50.41	37.91	5.14	14.91	37.25	41.12	34.68
X_6	23.88	39.17	40.70	20.32	9.58	2.01	56.30	51.25	57.29
X_7	28.26	31.26	38.61	27.66	9.66	8.48	44.08	59.08	52.91
X_8	20.61	39.82	30.92	15.14	0.50	4.68	64.25	59.68	64.40
X_9	18.76	45.00	26.32	34.62	5.16	19.03	46.62	49.84	54.65
X_{10}	27.75	23.30	20.55	17.54	12.63	19.73	54.71	64.09	59.72
X_{11}	32.21	23.15	36.32	0.54	17.23	2.46	67.25	59.62	62.22
X_{12}	30.13	35.90	33.79	6.66	11.61	13.62	63.21	52.49	52.59
X_{13}	30.58	37.52	31.02	7.22	0.10	3.61	62.20	62.38	65.37
X_{14}	23.19	47.68	36.51	19.97	7.60	1.84	56.84	44.72	61.65
X_{15}	37.49	41.22	32.20	12.51	4.33	3.85	50.00	54.45	63.95
X_{16}	33.78	37.94	31.69	10.68	3.53	7.01	56.54	58.53	61.30

(Source: Thirugnanakumar, 1991)

6.1.2.4. Mean performance

The mean performance for seed yield alone presented (Table 47). Among the lines it varied from 0.19 g (L_9) to 3.12 g (L_2) in S_1. Among the testers, it ranged from 0.51 g (T_1) to 2.22 g (T_2) in S_1. In S_2, it varied from 0.86 g (L_{20}) to 5.32 g (L_{23}) among the lines, whereas it ranged from 2.31 g (T_3) to 3.65 g (T_1) among the testers. In the pooled analysis, L_{13} registered maximum (3.97 g) and L_3 recorded minimum (1.05 g). Among the testers, it varied from 1.84 g (T_3) to 2.34 g (T_2). Among the 81 hybrids evaluated in the first season, $L_{10} \times T_1$ showed the minimum seed yield (0.45 g) and $L_{23} \times T_3$ recorded the maximum (9.45 g) followed by $L_{21} \times T_3$ (8.64 g) and $L_6 \times T_3$ (6.99 g). In S_2, a minimum of 1.25 g per plant seed yield was recorded by $L_1 \times T_3$ and a maximum of 8.50 g was registered by $L_{12} \times T_1$, which was closely followed by $L_{16} \times T_1$ (8.29 g) and $L_{13} \times T_2$ (7.84 g). In the pooled analysis, $L_{26} \times T_2$ and $L_1 \times T_3$ recorded the minimum seed yield (1.29 g) and $L_{23} \times T_3$ expressed maximum (6.98 g) followed by $L_6 \times T_3$ (5.87 g), $L_{13} \times T_2$ (5.42 g) and $L_{10} \times T_2$ (5.29 g).

Table 47 : Mean performance of parents and hybrids in individual and pooled over seasons – seed yield (g)

Lines	Mean			Testers								
				T_1			T_2			T_3		
	S_1	S_2	P	S_1	S_2	P	S_1	S_2	P	S_1	S_2	P
L_1	1.22	3.47	2.35	1.28	3.03	2.15	0.95	2.96	1.95	1.32	1.25	1.29
L_2	3.12	3.16	3.14	2.64	4.30	3.47	0.53	5.60	3.06	2.38	3.80	3.09
L_3	0.75	0.36	1.05	1.64	2.21	1.92	0.99	4.41	2.70	2.17	3.23	2.70
L_4	0.75	2.99	1.87	1.37	3.18	2.27	1.50	2.33	1.91	2.58	3.08	2.83
L_5	1.27	2.65	1.96	2.24	3.25	2.74	1.00	3.89	2.44	2.03	2.24	2.13
L_6	1.55	2.05	1.80	2.21	3.97	3.09	1.18	2.66	1.92	6.99	4.76	5.87
L_7	2.35	4.33	3.34	1.89	3.45	2.67	1.90	6.40	4.15	1.96	3.49	2.73
L_8	0.43	2.47	1.45	1.91	2.66	2.29	2.08	1.65	1.86	4.80	1.96	3.38
L_9	0.19	1.29	0.74	2.70	4.17	3.44	1.65	2.54	2.09	2.70	1.79	2.24
L_{10}	1.02	4.23	2.63	0.45	6.43	3.44	3.31	7.28	5.29	2.60	4.12	3.36
L_{11}	2.30	2.88	2.59	0.83	4.37	2.60	0.83	2.71	1.77	2.43	3.76	3.09
L_{12}	0.67	2.72	1.69	0.80	8.50	4.65	2.33	6.58	4.45	2.15	2.84	2.49
L_{13}	2.67	5.26	3.97	0.82	3.31	2.06	3.01	7.84	5.42	1.67	3.46	2.57
L_{14}	0.83	1.36	1.09	1.36	4.87	3.11	3.80	3.92	3.86	1.45	3.15	2.30
L_{15}	1.00	1.68	1.34	1.49	6.13	3.81	2.90	3.12	3.01	1.74	4.78	3.26
L_{16}	1.17	1.48	1.32	1.32	8.29	4.80	4.40	5.10	4.75	0.76	2.80	1.78
L_{17}	2.35	3.21	2.74	1.07	5.23	3.15	2.21	3.73	2.97	3.31	2.63	2.97
L_{18}	1.51	2.56	2.03	0.97	6.25	3.61	3.13	1.78	2.45	0.68	4.06	2.37
L_{19}	1.86	1.52	1.69	2.40	3.12	3.76	1.01	3.02	2.01	3.53	2.19	2.86
L_{20}	1.36	0.86	1.11	0.96	3.44	2.20	0.92	4.16	2.54	1.19	4.19	2.69
L_{21}	0.64	4.25	2.45	0.82	2.56	1.69	0.62	2.28	1.45	8.64	2.06	5.35
L_{22}	1.02	1.18	1.10	1.83	2.63	2.23	1.69	2.12	1.91	3.32	2.34	2.83
L_{23}	2.28	5.32	2.80	1.70	3.80	2.75	3.52	6.79	5.15	9.45	4.51	6.98
L_{24}	1.80	3.23	2.52	2.05	3.55	2.80	1.40	2.78	2.09	4.89	2.82	3.86
L_{25}	2.55	2.69	2.62	1.15	3.40	2.28	1.49	1.77	1.63	3.84	4.07	3.95
L_{26}	0.30	5.12	2.71	1.27	3.80	2.53	3.90	1.67	1.29	3.84	3.20	3.52
L_{27}	0.68	4.29	2.48	2.05	3.04	2.54	0.93	3.80	2.37	3.28	2.27	2.77
Tester Mean				**0.51**	**3.65**	**2.08**	**2.22**	**2.45**	**2.34**	**1.37**	**2.31**	**1.84**

Mean of the standard variety (Co.1) : S_1 = 2.14; S_2 = 1.95; P = 2.05
(Source: Thirugnanakumar, 1991)

6.1.2.5. Combining ability effects

Combining ability effects for seed yield alone presented (Table 48). Among the lines, L_{23} followed by L_6 and L_{21} in S_1; L_{12} followed by L_{10}

Table 48 : Combining ability effects of parents and hybrids in individual and pooled over seasons – seed yield (g)

Lines	gca			sca Testers								
					T_1			T_2			T_3	
	S_1	S_2	P	S_1	S_2	P	S_1	S_2	P	S_1	S_2	P
L_1	-1.00**	-1.30**	-1.15**	0.76**	0.14	0.45	0.09	0.45	0.27	-0.85**	-0.59	0.72*
L_2	-0.34*	0.86*	0.26	1.45**	-0.74	0.35	-0.99**	0.94	-0.03	-0.46	-0.20	0.32
L_3	-0.59**	-0.43	-0.51**	0.70**	-1.54**	-0.42	-0.28	1.03	0.38	-0.42	0.51	0.04
L_4	-0.37*	-0.85*	-0.61**	0.21	-0.15	0.03	0.01	-0.63	-0.31	-0.22	0.78	0.28
L_5	-0.43**	-0.58	-0.51**	1.15**	-0.35	0.40	-0.43	0.67	0.12	-0.72**	-0.32	-0.52
L_6	1.28**	0.08	0.68**	-0.59*	-0.30	-0.44	-1.95**	-1.23*	-1.59**	2.54**	1.53**	2.03**
L_7	-0.27	0.74*	0.23	0.63*	-1.47*	-0.42	0.31	1.86**	1.08**	-0.94**	-0.39	-0.66*
L_8	0.74**	-1.62**	-0.44*	-0.36	0.10	-0.13	-0.52*	-0.54	-0.53	0.88**	0.44	0.66*
L_9	0.16	-0.88*	-0.36	1.01**	0.86	0.94**	-0.37	-0.39	-0.38	-0.64*	-0.47	-0.56
L_{10}	-0.07	2.23**	1.08**	-1.01**	0.02	-0.50	1.51**	1.24*	1.38**	-0.50	-1.26*	-0.88**
L_{11}	-0.82**	-0.10	-0.46*	0.13	0.28	0.21	-0.20	-1.00	-0.60	0.07	0.72	0.39
L_{12}	-0.42**	2.26**	0.92**	-0.30	2.06**	0.88**	0.90**	0.50	0.70*	-0.60*	-2.56**	-1.58**
L_{13}	-0.35*	1.16**	0.40*	-0.36	-2.03**	-1.19**	1.50**	2.87**	2.18**	-1.14**	-0.84	-0.99**
L_{14}	0.02	0.27	0.14	-0.19	0.42	0.11	1.93**	-0.16	0.88**	-1.74**	-0.26	-0.99**
L_{15}	-0.15	0.96**	0.41*	0.11	0.98	0.54	1.18**	-1.65**	-0.24	-1.29**	0.67	-0.30
L_{16}	-0.03	1.68**	0.83**	-0.18	2.43**	1.12**	2.57**	-0.40	1.09**	-2.39**	-2.03**	-2.21**
L_{17}	0.01	0.15	0.08	-0.47	0.90	0.21	0.34	-0.23	0.06	0.13	-0.67	-0.27
L_{18}	-0.59**	0.32	-0.14	0.04	1.75**	0.90**	1.86**	-2.35**	-0.25	-1.90**	0.60	-0.65*
L_{19}	0.13	-0.93**	-0.40*	0.74**	-0.13	0.31	-0.97**	0.14	-0.42	0.23	-0.01	0.11
L_{20}	-1.16**	0.22	-0.47*	0.60*	-0.96	-0.18	0.22	0.13	0.17	-0.82**	0.83	0.01
L_{21}	1.17**	-1.41**	-0.12	-1.88**	-0.21	-1.05**	-2.1**	-0.12	-1.26**	4.29**	0.33	2.31**
L_{22}	0.09	-1.35**	-0.63**	0.21	-0.21	0.00	-0.26	-0.34	-0.30	0.05	0.55	0.30
L_{23}	2.70**	1.32**	2.01**	-2.53**	-1.70**	-2.12**	-1.04**	1.66**	0.31	3.57**	0.04	1.81**
L_{24}	0.50**	-0.66	-0.03	-0.07	0.03	-0.02	-1.05**	-0.37	-0.71*	1.12**	0.34	0.73*
L_{25}	-0.03	-0.63	-0.33	-0.35	-0.15	-0.25	-0.34	-1.41*	-0.88**	0.69**	1.56**	1.13**
L_{26}	-0.18	-0.82*	-0.50**	-0.08	0.44	0.18	-0.77**	-1.32*	-1.05**	0.85**	0.88	0.87**
L_{27}	-0.10	-0.67	-0.39*	0.62*	-0.47	0.08	-0.82**	0.67	-0.08	0.20	-0.20	0.00
Inter Mean				**-0.66****	**0.47****	**-0.09**	**-0.33****	**0.10**	**-0.11**	**0.99****	**-0.57****	**0.20****

$T(s_i)$ lines	0.15	0.34	0.18
$T(s_j)$ testers	0.05	0.11	0.06
$T(s_{ij})$ hybrids	0.6	0.58	0.32

* significant at P = 5% ** significant at P = 1%

(Source: Thirugnanakumar, 1991)

and L_{16} in S_2 and L_{23} followed by L_{10} and L_{12} in the pooled analysis expressed high positive and significant *gca* effects. In the case of testers, T_3 in S_1 and pooled analysis and T_1 in S_2 exhibited significant positive *gca* effects.

Among the hybrids, $L_{21} \times T_3$ followed by $L_{23} \times T_3$, $L_{16} \times T_2$ and $L_6 \times T_3$ in S_1, $L_{13} \times T_2$ followed by $L_{16} \times T_1$ and $L_{12} \times T_1$ in S_2 and $L_{21} \times T_3$ followed by $L_3 \times T_2$ and $L_6 \times T_3$ in pooled analysis exhibited highest positive and significant *sca* effects. Among the four cross combinations which had maximum positive significant *sca* effects in S_1, the cross combinations *viz.*, $L_{21} \times T_3$, $L_{23} \times T_3$ and $L_6 \times T_3$ had both the parents with high positive *gca* effects. It may indicate the involvement of additive and additive × additive gene effects. It is true for the two cross combinations *viz.*, $L_{16} \times T_1$ and $L_{12} \times T_1$ in S_2 and $L_6 \times T_3$ in the pooled analysis. On the other hand the cross *viz.*, $L_{13} \times T_2$ in S_1 and pooled analysis which had high positive *sca* effects had female parent with high positive *gca* effects and male parent with non-significant *gca* effects. Similarly, the cross combination *viz.*, $L_{21} \times T_3$ which had high positive *sca* effects had female parent with non-significant *gca* effect and male parent with significant positive *gca* effects. It is probably due to the contributions of dominant genes from one parent and recessive genes from other parent, resulting in high *sca* effects due to dominance and may also be due to epistatic effects. These cross combination promote the maximum expression of another, it may be a catalyst, a synergist, a stimulator and may be a complementary (Langham, 1961).

6.2. Diallel analysis

Breeding of plants and animals is an exercise for exploiting and manipulating the genetic system. Williams (1964) has suggested that refined and deliberate methods of breeding can be developed only from a more thorough knowledge and understanding of the genetic basis of traits. In the breeding programme, it is desirable to obtain information on the ability to select individuals which when combined together can give rise to superior progeny (Hayward, 1979). Hence, breeders have been giving importance for choice of parents for hybridization based upon the genetic potentiality rather than on their phenotypic performance. The genetic potentiality of parents is assessed based upon their mean performance, combining ability and their effects on hybrid performance. A number of methods are now available to assess the genetic potentiality of the parents. The combining ability analysis gives useful information regarding the selection of parents in terms of the performance of hybrids and in addition, it indicates the nature and magnitude of various types of gene action involved in the expression of the quantitative traits (Dhillon, 1975).

Different methods have been developed to estimate the combining ability of which diallel analysis is one of the systematic approach which offers genetic evaluation of parents under study and also permits the identification and selection potential of the crosses in early generation (Johnson, 1963).

6.2.1. Studies on diallel analysis

Diallel mating design is used to estimate the genetic components of variation and combining ability of inbreds in a series of crosses. The application of diallel cross, procedures for the analysis of polygenic traits in self-pollinated species was developed by Jinks (1954) and Hayman (1954a).

Diallel cross is a sophisticated application of Vilmorin's isolation principle or progeny test (Hayes and Immer, 1942). Danish geneticist, Schmidt (Lush, 1945; Wricke and Weber, 1986) first used the name diallel to designate a cross of two males with two females. The first written record of "diallel cross" as applied in plants was found in 1953 (Jinks and Hayman, 1953). The diallel cross was defined as all possible crosses among a group of parents including the parents themselves.

There would be n^2 families with n parents (Jinks and Hayman, 1953). The n^2 families have also been called as a "Complete diallel cross" (Crumpacker and Allard, 1962). If the reciprocal crosses are combined, making (n (n-1)/2) families the result is a "half diallel" (Morley-Jones, 1965). A "modified diallel" is one in which the parents are not included (Griffing, 1956a). A partial diallel includes fewer than the (n (n-1)/2) crosses (Kempthorne and Curnow, 1961).

There are four basic designs and analyses. They are:

1. Analysis of general and specific combining ability often referred to as Griffing's analysis (Griffing, 1956b).
2. Analysis of array variances and covariances, often referred to as Hayman and Jinks analysis (Jinks and Hayman, 1953; Hayman, 1954b; Jinks, 1954).

3. Analysis of additive and dominance effects often referred to as Gardner and Eberhart's analysis (Gardner and Eberhart, 1966; Eberhart and Gardner, 1966).
4. Partial diallel (Gilbert, 1958; Kempthorne and Curnow, 1961).

6.2.1.1. Griffing's analysis

Griffing's (1956b) proposed a diallel technique for determining the combining ability of lines and characterising the nature and extent of gene action in both plants and animals. His approach has also been adapted to assess competition (Ames-Gottfred and Christe, 1989). Since its formulation, Griffings' analysis has been widely used by plant breeders (Buiatti *et al.*, 1974; Gomaa Gibral *et al.*, 1982; Williams and Windham, 1988; Krueger *et al.*, 1989). Wright (1985) describes the three levels of the analysis and discusses the required assumptions at each level. Griffing's analysis allows the option to test for fixed (model I) or random (model II) effects.

Griffing (1956b) proposed four methods of diallel crossing:

Method 1 (Full diallel) : The parents, F_1's and reciprocals included (p^2 total entries, where p is the number of parents)

Method 2 (Half diallel) : Parents and F_1's included, but no reciprocals (p (p+1) /2 total entries)

Method 3 : F_1's and reciprocals included, but no parents (p^2-p total entries)

Method 4 : F_1's included, but no reciprocals or parents (p(p-1)/2 total entries).

Griffing (1956b) formulated detailed analysis of general and specific combining ability in relation to diallel crossing systems. Eight different types of analysis were presented by considering four different diallel crossing system together with two alternative assumptions (fixed effect i.e., model 1 and random effect i.e., model II) with regard to the sampling nature of the experimental material.

Griffing's analysis on combining ability requires no genetic assumptions (Wright, 1985) and has been shown to convey reliable information on the combining potential of parents (Gill *et al.*, 1977; Bhullar *et al.*, 1979; Nienhuis and Singh, 1986). Once identified, the best parental combiners can be crossed to identify optimal hybrid combinations or hybridized with the intent of selecting promising genotypes within the segregating generation. In recurrent selection schemes parents possessing high combining ability can be crossed with one another in an attempt to accumulate desirable alleles within a base population.

Sprague and Tatum (1942) defined general combining ability (*gca*) as "The average performance of a line in hybrid combinations" and specific combining ability (*sca*) as "those cases in which certain combinations do relatively better on worse than would be expected on the basis of the average performance of the lines involved".

Rojas and Sprague (1952) and Griffing (1956a) stated that the *gca* includes the additive genetic portion, while *sca* includes non-additive genetic portion of the total variation. Hayman (1957) found that in the absence of epistasis, *gca* was composed of both additive and dominance effects. On the other hand, when epistasis was present, both the combining abilities contained epistatic portions. In *gca,* this portion is average epistatic effect in the corresponding array of that parent, while *sca* relates more directly to epistasis in a particular cross.

A relatively large *gca* / *sca* variance ratio suggests the importance of additive gene effects, and a low ratio implies the presence of dominance and/or epistatic gene effects (Griffing 1956a; Bhullar *et al.,* 1979). It should be noted that if additive x additive effects are present, the *gca* component will also contain some of those effects in addition to additive effects (Wassimi *et al.,* 1986). Where *sca* is small relative to the *gca,* performance of single cross progeny can be predicted on the basis of the *gca* of the parents. For inbred parents, the closer the following equations are to unity, the greater the predictability based on general combining ability (Baker, 1978).

The choice of a fixed or random effects model will determine if individual *gca* and *sca* effects for parents and crosses respectively should be estimated or if the variance of these effects should be estimated. When a limited number of parents are evaluated and the selection of these parents cannot be considered random, then a fixed model is recommended.

The choice of the Griffing's method will depend on the researcher's preference and on the characteristic of the crop and trait under evaluation. For example, when the trait is influenced by maternal effects or cytoplasmic inheritance, method 1 and method 3 with modifications (Borges, 1987) may be used. When reciprocal effects are absent, the means of the reciprocal hybrids can be combined for analysis. Where the variance components are of interest, Pooni *et al.* (1984) suggests the use of method 1, since it was shown to give a more accurate constant variance estimate than the other methods, in most situations.

Bray (1971) and Wright (1985) pointed out three possible levels of diallel analysis, *viz.*, A) Estimation of general and specific combining ability, B) Estimation of genetic variance components, and C) A complete genetic analysis.

6.2.1.2. Hayman and Jinks analysis

Jinks and Hayman (1953); Hayman (1954a, 1954b) and Jinks (1954) developed a method of analysis that has been widely used for evaluating the mode of inheritance (Radwan and El-Zahab, 1974; Dwivedi *et al.*, 1980; Koevering *et al.*, 1987; Powell, 1988). This analysis is based on a model that, for any one locus *i* with two alleles, the difference between two homozygotes is 2di. The difference between the heterozygote value is h_i, while the frequency of the two alleles is Ui and Vi.

The parents and all possible F_1 progenies are evaluated for the trait(s) of interest. All offspring of one parent is called as an "array" that is all crosses with that particular parent. Six kinds of variances and covariances are calculated and to properly interpret these, the diallel cross must meet the following assumptions.

1. Diploid segregation
2. Homozygous parents
3. No difference between reciprocal crosses
4. No epistasis
5. No multiple alleles
6. Genes distributed independently among the parents

The six kinds of variances and covariances are:

V_p - Variance among the parents

V_r - The variance among family (F_1 + reciprocal) means within an array

V_r – Mean value of V_r over all arrays

V_r' – Variance among the means of arrays

W_r - The covariance between families within the ith array and their non recurrent parent

W_r' - Mean value of W_r over all arrays

By adopting these, we can estimate the following five parameters

$\hat{E}$ = Error /environmental variance

$\hat{D}$ = Component of variation due to additive effect of genes

$\hat{H}_1$ = Component of variation due to dominance effect of genes

$\hat{H}_2$ = The proportion of dominance variance due to positive and negative effect of genes

$\hat{F}$ = The mean value of F_r over arrays, F_r being covariance of additive effect in the rth array.

$\hat{h}^2$ = Dominance effect expressed as the algebraic sum over all loci in heterozygous phase in all crosses.

In some cases, the variance among parents may be different than the variance among F_1 families. For example inbred lines of maize may have environmental variances different from those of F_1 families. When the parent and F_1 variances are unequal, the above formulae have to be modified to include both the variances (Crumpacker and Allard, 1962).

The estimates of and indicate the relative amounts of additive variance and dominance variance among the crosses. The estimate of will be positive when the dominant allele is the more frequent and negative when the recessive allele is more frequent. If the frequency of dominant and recessive alleles are equal then the estimate of F=0.

From the variances and covariances, the following relations can be derived.

1. The regression W_r=¼D-¼H_1+bV_r. The intersection of the line with the Wr axis (i.e., V_r = 0) is represented as W_r = ¼D - ¼H_1. The intersection can be used to estimate the relative level of dominance. If H =D, W_{ri} = 0, and the intersection will be below the point of origin and if *h*<*d*, the intersection will be above the origin.

2. For one locus, if the parent is homozygous dominant, the values for W_r and Vr will be less than that for the parent, which is homozygous recessive. The parent with the greater number of dominant alleles have a smaller Wr and Vr, and thus will be located on the regression line closer to the origin, the parent with the greater number of recessive alleles. By plotting the values of Wr / Vr for each parent (on a graph) one can rank them according to the relative numbers of loci carrying dominant versus recessive alleles.

3. The ratio of Vr / Wr can be used to assess the relative values of h and d. For example, if d = h, this ratio will equal to 1. If $h<d$, incomplete dominance, the ratio will be less than one.

4. The value for the parents with dominant alleles at all loci or with the recessive alleles at all loci can be estimated. These two estimates would represent the limits to selection within the material being studied.

5. To calculate these limits, it should be noted that the regression of Wr on Vr has its limits at the intersection with a parabola, whose form is $Wr^2 = (Vr)(Vp)$. From the equations for the parabola and for the regression, the points of intersection can be calculated.

6. The correlation between the parental order of dominance (Vr + Wr) and the parental mean (Yr) will be positive if recessive alleles contribute to an increase in mean, and negative if recessive alleles contribute to a lower mean. A low and non-significant correlation indicates that recessive alleles contribute both positive and negative effects. If the correlation is high (close to + 1 or –1) then from the above points of intersection, the means for the parents at the limits can be estimated.

6.2.2. Methods

The statistical analysis for the full diallel analysis based on Griffing (1956b) model I (fixed) method 1 procedure is given below.

Analysis of variance for RBD

Source of variation	Degrees of freedom	M.S.	Expectations of mean squares
Replication	r-1	M_r	$s^2e + gs^2r$
Genotype	g-1	M_g	$s^2e + rs^2g$
Error	(r-1) (g-1)	M_{rg}	s^2e

(Source: Griffing, 1956b)

where,

r = number of replications

g = number of genotypes

Combining ability analysis

The procedure outlined by Griffing (1956b) for Model I (fixed), Method 1 was considered appropriate for this study it involved the parents, F_1 of both direct and reciprocal crosses.

a) Analysis of variance for combining ability

Source	Degrees of freedom	Sum of squares	Mean squares	Expectation of mean squares
GCA	P-1	S_g	M_g	$\sigma^2e + 2P\,[1/(P\text{-}1)\Sigma_i g_i^2$
SCA	[P(P-1)] / 2	S_s	M_s	$\sigma^2e + [2/P(P\text{-}1)]\Sigma_i\Sigma_j s_{ij}^2$
Reciprocal effects	[P(p-1)] / 2	S_r	M_r	$\sigma^2e + 2\,[2/P(P\text{-}1)]\Sigma_i\Sigma_{<j} r_{ij}^2$
Error	m	S_e	Me′	σ^2e

(Source: Griffing, 1956b)

where,

P = Number of parents

$$S_g = \frac{1}{2P} \sum_i (Xi. + X.j)^2 - \frac{2}{P^2} X^2 ....$$

$$S_s = \frac{1}{2} \Sigma_i\Sigma_j X_{ij}(X_{ij} + X_{ji}) - \frac{1}{2P} \Sigma_i (X._j + X._i)^2 + \frac{1}{P^2} X^2 ...$$

$$S_r = \frac{1}{2} \Sigma_i\Sigma_{<ij}(X_{ij} - .X_{ji})^2 ...$$

Me′ = Me / r, where, Me is the EMS from ANOVA for RBD.

F_{gca} = Mg/Me′, F_{sca} = Ms /Me′, F_r = Mr /Me′

General combining ability effects, specific combining ability effects and reciprocal combining ability effects were estimated as follows.

$g_i = (1/\ 2P) \times (x_{i.} + x_{.i}) - (1/P^2) \times ....$

$s_{ij} = (1/2)\ (x_{ij} + x_{ji}) - (1/2P)\ (x_{i.} + x_{.j} + x_{j.} + x_{.i}) + (1/P^2) \times ....$

$r_{ij} = (1/2)\ (x_{ij} - x_{ji})$

b) Standard error of estimates

The square root of the variance gives the standard error of that estimate. For this, the variance of different effects were calculated as follows.

$$\text{Variance } (X_{ij}) = \sigma^2_e = Me'$$

$$\text{Variance } (g_i) = \frac{P-1}{2P^2}\sigma^2_e$$

$$\text{Variance } (s_{ij}) = \frac{1}{2P^2}(P^2 - 2P + 2)\sigma^2_e$$

$$\text{Variance } (r_{ij}) = \frac{1}{2}\sigma^2_e$$

c) Critical difference of estimates

The critical difference to compare any two similar estimates was obtained as a product of 't' value against error degrees of freedom and standard error of difference. The square root of the variance gives the standard error of differences.

Variance $(g_i - g_j) = (1\ /\ P)\ \sigma^2_e$ for two *gca* estimates

Variance $(s_{ij} - s_{ik}) = [(P\ -1)/P]\ \sigma^2_e$ for two *sca* estimates having one parent in common.

Variance $(s_{ij} - s_{kl}) = [(P-2)\ /\ P]\ \sigma^2_e$ for two *sca* estimates with no parent in common

Variance $(r_{ij} - r_{kl}) = \sigma^2_e$ for a pair of reciprocal effects.

d) Scoring based on combining ability effects

The parents or cross combinations which showed significant positive *gca* or *sca* or *rca* effects were given the score +1. The parents or cross combinations which recorded significantly negative *gca* effects were given the score -1. The parents or cross combinations which registered non-significant *gca* or *sca* or *rca* effects were given the score 0. For days to seedling emergence, days to first flowering number of first fruiting node height of first fruiting node, plant height and days to first picking negative significant *gca* or *sca* or *rca* effects were given the score +1. The genotype, which exhibited a total score more than +1 was considered as good combiner. The genotype, which scored a total score of -1 was considered as poor combiner. The genotype, which scored a total score of 0, was considered as an average combiner.

e) Simple correlation coefficients

Simple correlation coefficients were estimated among the mean performance of parents and their *gca* effects, as well as among the mean performance of the hybrids and the *sca* / *rca* effects of their cross combinations.

$$r\ (1.2) = \frac{\text{Covariance between 1 and 2}}{(\text{Variance of 1 x variance of 2})^{1/2}}$$

1 = Mean performance

2 = Effects

The significance of these correlation coefficients was tested by referring to the table given by Snedecor and Cochran (1961).

Graphic analysis

The graphic analysis proposed by Jinks and Hayman (1953) was followed. The variance and covariance listed below were calculated as follows.

V_r = Variance of the off spring of rth parental array

W_r = Covariance between the non - recurring parents and the offspring of rth array

W_r' = Covariance between array means and the offspring of rth array

V_0L_0 = Variance of the parents

V_0L_1 = Variance of the mean of arrays

V_1L_1 = Mean variance of the arrays

W_0L_{01} = Mean covariance between the parents and the arrays

$(M_{L1\ c}\, M_{Lo})^2$ = The square of difference between mean of the parents and mean of their n^2 progenies.

E = Expected environmental component of variance

Utilizing the regression values, V_r W_r; W_r W_r' and Y_r, $(V_r + W_r)$ graphs were drawn for the characters studied. The limiting parabola for V_r W_r graph was constructed with the formula, $W_r^2 = V_r \times V_P$. The correlation co-efficient 'r' for Y_r, $(V_r + W_r)$ graph was also estimated.

Genetic analysis

The genetic components *viz.*, $\hat{D}$, $\hat{F}$, $\hat{H}_1$, $\hat{H}_2$, $\hat{h}^2$ and $\hat{E}$ were estimated as described by Hayman (1954a) using the second degree statistics and error mean square as follows.

$$\hat{D} = V_{OLO} - \hat{E}$$

$$\hat{F} = 2V_{OLO} - 4W_{OLO1} - 2(n-2)\hat{E}/n$$

$$\hat{H}_1 = V_{OLO} - 4W_{OLO1} - 4V_{1L1} - (3n-2)\hat{E}/n$$

$$\hat{H}2 = 4V_{1L1} - 4V_{OL1} - 2\hat{E}$$

$$\hat{h}^2 = 4(M_{L1} - M_{LO})^2 - 4(n-1)\hat{E}/n^2$$

and $\hat{E}$ = Expected environmental variation obtained as sum of squares for blocks + sum of squares for replication and progeny / degrees of freedom for both x number of replications.

where

$\hat{D}$ = Component of Variation due to additive effect of genes

$\hat{F}$ = The mean value of F_r over arrays, F_r, being covariance of additive effect in the rth array.

$\hat{H}_1$ = Component of variation due to the dominance effect of the genes.

$\hat{H}_2$ = The proportion of dominance variance due to positive and negative effect of genes

$\hat{h}^2$ = Dominance effect expressed as the algebraic sum over all loci in heterozygous phase in all crosses.

The standard error for the above estimates were calculated using the equation $S^2 = \frac{1}{2}$ (var $(W_r - V_r)$ as the common multiplier and the term of the main diagonal of covariance matrix as corresponding multipliers.

These genetic parameters thus calculated were employed for the computation of proportional values mentioned below.

Mean degree of dominance $= (H_1/D)^{½}$

The proportion of genes with positive and negative effects in the parents $= H_2 / 4H_1$

The proportion of dominant and recessive genes in the parents $= \frac{[4DH_1)^{1/2} + F]}{[4DH_1)^{1/2} - F]}$

Number of groups of genes which control the character and exhibit dominance k $= \frac{h^2}{H^2}$

Heritability in the narrow sense h^2 (ns) $= \frac{(1/2)D + (1/2)\,H_1 - (1/2)H_2 - (1/2)F}{(½)D + (½)\,H_1 - (¼)H_2 - (½)F + E}$

Test of hypothesis

1. The validity of the hypothesis as postulated by Hayman (1954a) was tested by analysis of variance of V_r, W_r. The homogeneity of V_r, W_r would indicate the validity and it was tested as follows.

   $$t^2 = \frac{(n-2)}{4} x \frac{(Var.V_r - Var.W_r)^2}{(Var.V_r + Var.W_r) - Cov^2(V_r, W_r)}$$

   with (n-2) degrees of freedom. Significant t^2 indicates the failure of the hypothesis.

2. The regression (b) of covariance (Wr) on variance (Vr) and its S.E was calculated as

   $$b = \frac{Cov.(W_r, V_r)}{Var.V_r}$$

   where,

   $$Cov(W_r, Vr) = \left[\sum VrWr - \frac{\sum Vr \sum Wr}{n}\right] \Big/ (n-1)$$

$$\text{S.E.(b) for Vr, Wr} = \left[\frac{\text{Var.Wr.} - \text{bCov.WrVr}}{\text{Var.Vr}(n-2)}\right]^{1/2}$$

$$\text{S.E.(b) for Wr, Wr}' = \left[\frac{\text{Var.Wr}',-\text{bCov.WrWr}}{\text{Var.Wr}(n-2)}\right]^{1/2}$$

Significance of deviation of b values from zero and one was tested as H_0 : b = 0 = b - 0 / SE (b) and H0 : b =1 = 1 - b/SE (b)

Both the values were tested by 't' for (n-2) degrees of freedom. The point of intersection of the regression line with Wr ordinate i.e. 'a' is obtained by the following equation.

$a = \overline{Wr} - b\overline{Vr}$

6.2.3. Example

Eswaran (2007) made full diallel analysis in bhendi with six parents, the evaluated 30 hybrids along with their six parents to assess their genetic potentiality. He recorded data on 10 earliness and fruit yield characters. The data were analyzed using Griffing's (1956b) methods of combining ability analysis. The result on fruit yield alone is discussed here under.

6.2.3.1. Analysis of variance (ANOVA)

The ANOVA (Table 49) revealed that the parents, direct as well as their reciprocal hybrids differed among themselves for all the six characters studied. This indicated the presence of high genetic variability in the reference population. Therefore, further analysis was appropriate.

Table 49 : ANOVA for fruit yield and yield components in Okra

S. No.	Characters	df	MSS	'F' value
1.	Days to seedling emergence	35	0.38	10.56**
2.	Days to first flowering	35	15.86	11.79**
3.	Number of branches per plant	35	4.42	49.29**
4.	Number of first fruiting node	35	5.78	85.91**
5.	Height of first fruiting node	35	67.01	18.98**
6.	Plant height	35	354.80	12.43**
7.	Days to first picking	35	27.52	40.29**
8.	Number of fruits per plant	35	72.99	31.53**
9.	Fruit weight	35	13.82	10.45**
10.	Fruit yield per plant	35	44474.89	207.76**

** - Significant at 1 per cent level (Source: Eswaran, 2007)

6.2.3.2. Mean performance

Fruit yield per plant was maximum with parent Arka Anamika (AA) (Table 50). Among the cross combinations the hybrids Parbani Kranti × Arka Anamika and its reciprocal cross showed maximum fruit yield per plant.

Table 50 : Mean performance of parents and their hybrids for fruit yield per plant

Parents	AA	PK	PS	EC-112112	EC-305626	IC-128076
AA	**490.35**	685.31	563.27	358.39	377.04	391.12
PK	742.12	**441.82**	318.26	617.84	648.58	446.71
PS	429.29	298.84	**348.57**	438.61	499.23	328.73
EC-112112	394.73	672.61	443.23	**402.58**	360.75	511.03
EC-305626	403.13	682.83	369.23	343.63	**357.84**	319.74
IC-128076	385.99	449.48	318.91	458.24	373.54	**314.20**

Diagonal values indicate parental mean
General mean = 444.05
CD at 1 per cent level = 31.41± 8.44
(Source: Eswaran, 2007)

6.2.3.3. Combining ability variance

The mean sum of squares for combining ability revealed that both GCA and SCA variances were significant for all the characters studied by Eswaran (2007) (Table 51). This indicated the importance of both additive and non-additive genetic variances in the inheritance of the characters studied. The RCA variance was also significant for all the characters studied. The ratio of GCA/SCA was more than unity for all the characters studied except fruit weight. The ratio of GCA/RCA was also more than unity for all the characters except number of first fruiting node. This indicated the opulence of additive genetic variance in the improvement of the characters studied. The available additive genetic variance could well be exploited by resorting to simple pure like selection and/or pedigree breeding, whereas the non-additive genetic variance could well be exploited in later generations or by resorting to hybrid breeding. Both the variances could be exploited simultaneously by resorting to population improvement programme.

Table 51 : Estimates of variance for combining ability

Source	Mean sum of squares									
	Days to seedling emergence	Number of first fruiting node	Height of first fruiting node	Number of branches per plant	Plant height	Days to first flowering	Days to first picking	Number of fruits per plant	Fruit weight	Fruit yield per plant
GCA	0.33**	1.62**	53.59**	3.08**	377.72**	12.89**	24.11**	84.01**	5.62**	39321.00**
SCA	0.07**	1.00**	28.90**	1.98**	74.82**	5.47**	10.31**	24.25**	7.45**	19792.67**
RCA	0.11**	2.94**	5.37**	0. 42**	75.22**	2.54**	3.06**	4.52**	1.43**	1691.81**
GCA/SCA	4.71	1.62	1.85	1.55	5.04	2.35	2.33	3.46	0.754	1.99
GCA/RCA	3.00	0.55	9.98	7.33	5.02	5.07	7.88	18.59	3.93	23.24

** Significant at 1 per cent level (Source: Eswaran, 2007)

6.2.3.4. Combining ability effects

The success of any plant breeding programme largely depends on the correct choice of good parents. High mean value was the main criterion among the breeders for a long time. Gilbert (1958) suggested that the parents with good *per se* performance would result in better genotypes. Further, the parents having high *gca* effects could be useful since the *gca* effect is due to additive gene action and it is fixable (Sprague and Tatum, 1942). Dhillon (1975) reported that the combining ability of parents gives useful information on the choice of parent in terms of expected performance of the hybrids and their progenies. A number of workers and advocated *gca* effects, to critically analyze the parents for their ability to transmit their superior performance to their progenies. The *gca* effect is a value derived from the general mean of hybrids involving all the parents. The *gca* effects of the parents may be positive or negative. Simmonds (1979) pointed out that the *gca* values were relative and dependent on the mean of the chosen material. It is better to choose parents possessing significant *gca* effects or nearly based on mean performance. This assumption is based on the principle that *gca* effect reflects additive gene action. Sometimes, the immediate hybrid may not perform well despite both the parents possessing high *gca* effects for a trait due to the interaction of the parental *gca* effects which may cause distortions on expectations. The reverse trend may also happen with low performing parents showing high hybrid values than expected. The interaction is measured by *sca* effects of the hybrids. The specific combining ability is the deviation from the performance predicted based on the basis of general combining ability (Allard, 1960). According to Sprague and Tatum (1942), the specific combining ability is controlled by non-additive gene action. The *sca* effect is important for the evaluation of the hybrids.

Eswaran (2007) evaluated the combining ability effects for many characters in Okra. The data pertaining to fruit yield alone is discussed here under. The parents namely Parbhani Kranti, Arka Anamika, EC-112112 and EC-30562f were adjudged as good general combiners. The *sca* effects were positive and significant with Arka Anamika × Parbhani Kranti, Parbhani Kranti × EC-305626 and Parbhani Kranti × EC-112112. All these three cross combinations had both the parents with positive significant *gca* effects. It may indicate the presence of additive and additive × additive type of gene effects. The *rca* effects were positive and significant with Pusa Sawani × Arka Anamika; EC-305626 × Pusa Sawani and EC 305626 × EC 112112 (Table 52).

The first cross involved female parent with negative significant *gca* effect and male parent with positive significant *gca* effect. The reverse was true for the second cross. This may indicate the presence of dominance effect between the *gca* effects of the parents. The third cross had both the parents with positive significant *gca* effects. It may indicate the presence of additive and additive × additive type of gene effects.

Table 52 : Estimates of combining ability effects for fruit yield per plant in Okra

General combining ability (gca), specific combining ability (sca) and reciprocal combining ability (rca) effect

Parents	AA	PK	PS	EC-112112	EC-305626	IC-128076
AA	31.67**	104.65**	62.34**	-105.67**	-66.24**	-35.97**
PK	26.39**	64.14**	-176.65**	81.74**	98.12**	-29.37**
PS	66.99**	9.71	-51.99**	32.56**	41.77**	-8.52
EC-112112	-48.17**	8.56	-2.31	6.30**	-78.56**	39.47**
EC-305626	-43.05**	17.13**	44.99**	27.39**	9.60**	-18.08**
IC-128076	2.56	-1.38	4.91	-28.41**	-26.90**	-59.72**

S.E. for gca = 2.23   * Significant at 5 per cent level

S.E. for sca = 5.08   ** Significant at 1 per cent level

S.E. for rca = 5.97

Diagonal - General combining ability (gca) effects (Source: Eswaran, 2007)

Upper diagonal - Specific combining ability (sca) effects

Lower diagonal - Reciprocal combining ability (rca) effects

6.2.3.5. Ranking based on combining ability effects

Eswaran (2007) identified the following parents *viz.*, Parbani Kranti, EC 112112, EC 305626 and Arka Anamika as good general combiners based on over all general combining ability effects (Table 53). The following cross combinations were identified as good specific combiners based on the overall specific combining ability effects of the direct crosses. They are: Arka Anamika × Parbani Kranti; Pusa Sawani × EC 112112; Arka Anamika × Pusa Sawani; Parbani Kranti × EC 112112 and Pusa Sawani × EC 305626 (Table 54).

Table 53 : Scoring based on *gca* effects for all the ten characters

S. No.	Characters	Arka Anamika	Parbhani Kranti	Pusa Sawani	EC-112112	EC-305626	IC-128076
1.	Days to seedling emergence*	0	+1	-1	+1	0	-1
2.	Days to first flowering*	+1	+1	0	0	+1	-1
3.	Number of branches per plant	+1	+1	-1	+1	-1	-1
4.	Number of first fruiting node*	-1	0	-1	+1	+1	+1
5.	Height of first fruiting node*	-1	+1	+1	0	+1	-1
6.	Plant height*	+1	-1	-1	+1	+1	-1
7.	Days to first picking*	+1	+1	-1	-1	0	-1
8.	Number of fruits per plant	+1	+1	-1	+1	0	-1
9.	Fruit weight	0	+1	-1	0	+1	0
10.	Fruit yield per plant	+1	+1	-1	+1	+1	-1
	Total score	+4	+7	-7	+5	+5	-7

+1 = Positively significant -1 = Negatively significant 0 = Non-significant.
* - Negative significant effect taken as +1 and vice-versa.
(Source: Eswaran, 2007)

6.2.3.6. Relationship between per se performance of parents and their gca effects

The contribution of individual lines to hybrid performance was accomplished by comparing the general combining ability effects. The choice of parents based on *gca* effects should be continuously done since a parent possessing favourable genes for one trait may also be possessing genes governing unfavourable expression for some other traits. Only those parents which possess favourable *gca* effects for different characters should be used in crossing programme, so that segregation can be obtained with the superior performance for all the traits through recombination.

Eswaran (2007) recorded maximum fruit yield per plant with the parent namely Arka Anamika. This parent was also identified as good general combiner for days to first flowering, number of branches per plant, plant height at maturity, days to first picking, number of fruits per plant and fruit yield per plant. The parent namely, Arka Anamika has also registered higher mean performance for many of the fruit yield and its component characters. Thus, Eswaran (2007) observed a good agreement between *per se* performance and *gca* effects (Table 55). The high *per se* performance coupled with high *gca* effects in Arka Anamika indicated that it has enormous amount of additive genetic variability for fruit yield and its component characters. A simple selection in Arka Anamika may give some useful high yielding lines.

Table 54 : Scoring based on *sca* effects for fruit yield and its component characters in Okra

S. No.	Cross combinations	Days to seedling emergence	Days to first flowering	Number of branches per plant	Number of first fruiting node	Height of first fruiting node	Plant height	Days to first picking	Number of fruits per plant	Fruit weight	Fruit yield per plant	Total score
1.	AA x PK	0	+1	+1	+1	+1	+1	+1	+1	0	+1	+8
2.	AA x PS	+1	+1	+1	-1	-1	-1	+1	+1	0	+1	+3
3.	AA x EC-112112	-1	-1	-1	0	-1	+1	-1	-1	-1	-1	-7
4.	AA x EC-305626	+1	0	-1	+1	+1	+1	+1	-1	-1	-1	+1
5.	AA x IC-128076	-1	0	-1	0	-1	+1	0	+1	-1	-1	-3
6.	PK x PS	0	0	-1	0	+1	0	-1	-1	-1	-1	-4
7.	PK x EC-112112	0	-1	+1	+1	-1	0	+1	+1	+1	+1	+4
8.	PK x EC-305626	0	0	+1	0	0	-1	+1	+1	+1	+1	+4
9.	PK x IC-128076	0	-1	+1	+1	0	-1	-1	-1	-1	-1	-4
10.	PS x EC-112112	0	+1	-1	+1	+1	-1	+1	+1	+1	+1	+5
11.	PS x EC-305626	0	0	+1	0	0	0	0	+1	+1	+1	+4
12.	PS x IC-128076	+1	0	-1	-1	-1	0	-1	-1	0	0	-4
13.	EC-112112 x EC-305626	0	-1	-1	+1	-1	0	-1	-1	0	-1	-5
14.	EC-112112 x IC-128076	-1	-1	+1	-1	+1	0	-1	0	+1	+1	0
15.	EC-305626 x IC-128076	+1	0	+1	0	-1	-1	+1	0	0	-1	0

* Significant at 5 per cent level; ** Significant at 1 per cent level (Source: Eswaran, 2007)

The *gca* effects itself is considered to be the presence of a large number of favourable genes in parents for the traits concerned (Simmonds, 1979). As Arka Anamika has additive genetic variability, its ability to transmit desirable characters to the progeny could be predicted on the basis of its phenotypic performance. For an autogamous crop like Okra, additive genetic variability could be efficiently used by any conventional breeding programme which involves hybridization and selection. It mainly involves crossing of two or more diverse genotypes and then selecting in the segregating generations to fix the additive genetic variance.

Table 55 : Relationship between *per se* performance and *gca* effects

S No.	Characters	*Per se* performance	*gca* effect	Common parent	Correlation coefficient (r)
1.	Days to seedling emergence	PK EC-112112	PK EC-112112	PK EC-112112	0.92**
2.	Days to first flowering	PK -	PK AA	PK -	-0.01*
3.	Number of branches per plant	AA PS EC-112112	AA PK EC-112112	AA - EC-112112	0.41
4.	Number of first fruiting node	PK EC-112112 EC-305626	EC-305626 IC-128076 -	EC-305626 - -	0.41
5.	Height of first fruiting node	EC-112112 PK PS	PK PS EC-305626	PK PS	0.86**
6.	Plant height	EC-112112 EC-305626 PS	EC-112112 EC-305626 AA	EC-112112 EC-305626 -	0.93**
7.	Days to first picking	PK EC-112112 -	PK AA -	PK - -	0.237
8.	Number of fruits per plant	PK EC-112112 AA	PK AA EC-112112	PK AA EC-112112	0.83*
9.	Fruit weight	AA EC-305626 IC-128076	PK EC-305626 -	- EC-305626 -	0.32
10.	Fruit yield per plant	AA -	PK AA	AA -	0.82**

(Source: Eswaran, 2007)

6.2.3.7. Relationship between per se performance of hybrids and their sca effects

Riccharia and Singh (1983) stressed that the selection criteria for good cross is that it should have high *per se* performance and *sca* effects. Eswaran (2007) observed in his studies that most of the cross combinations which exhibited high *sca* effects had both the parents with high *gca* effects in Okra. However, in some crosses he noticed that at least one of the parent of the cross combinations which exhibited high *sca* effects had high *gca* effects. He has also recorded that the cross combinations with non-significant *sca* effects had parents with significant *gca* effects. It indicated that the parents with high general combining ability need not be good specific combiners (Singh *et al.*, 1983).

6.3. Triallel analysis

Three-way cross hybrids has been thought of as a viable alternative to single cross hybrids. Three-way cross hybrids have been fairly successful in sorghum (Walsh and Atkings, 1973) and cotton (Shroff *et al.*, 1983). It is opined that three way cross hybrids could be desirable under stress condition, since they are phenotypically stable than single cross hybrids (Weatherspoon, 1970).

6.3.1. Studies on triallel analysis

Triallel analysis provides the informations on all type of gene actions, selection methods suitable for segregating generations and the order of the parents in cross combinations for maximum range of segregation and recombinations and to exploit the heterosis at maximum extent possible (Rawlings and Cockerham, 1962). Triallel analysis in several crops namely maize, wheat, barley, sorghum, bajra, rice, groundnut and sesamum were carried out and the literature on these above crops are presented.

6.3.1.1. Maize

Weatherspoon (1970) compared single, three-way and double cross analysis in maize. He observed that the average and range of grain yield of single crosses were more than that of three-way crosses which was in turn more than that of double crosses. The results indicated that average yield superiority of single crosses over three-way crosses and three-way over double crosses.

Wright *et al.* (1971) made all possible single and three-way crosses from 60 inbred lines in maize. It was not possible to obtain realistic estimate of epistatic components although significant effects were evident in the analysis of variance. The estimate of additive genetic variance was significant for all traits in both the estimation procedures. Non-additive component accounted for only a small portion of the total genetic variances. The frequency of significant interactions with environments was highest for additive than dominance effects.

Ponnuswamy (1972) showed the relative importance of general and specific combining ability effects to triallel analysis in maize and its importance in choosing parents as well as order in which the parents are to be combined. Ponnuswamy *et al.* (1974) studied order effects, combining ability and gene action in maize. Absence of all types of epistasis except additive x additive for all characters was observed.

Giridharan (1986) studied six maize inbreds and their 15 single, 60 three way and 45 double cross hybrids for eleven cob characters for magnitude of variation, combining ability effects, parent order and gene action by subjecting to diallel, triallel and quadriallel analysis. Single cross hybrids were superior to three way which in turn, superior to double cross hybrids and all types of hybrids superior to parental lines for five characters *viz.*, husk, cob, grain, pith weight and cob length. Inbreds UMI 261 and UMI 180 were good combiners for seven cob characters and the single cross hybrid UMI 88 / UMI 261was the best cross based on combining ability effects and phenotypic means. Inbred UMI 61 was the best for use as one of the grand parent as well as third parent in three-way cross hybrids. The parent order affected the estimates of various effects and hybrid means. The different general line effect and two line specific effect were helpful in tracing the cross combination with high three line specific effect and high mean yielding hybrids for various cob characters.

6.3.1.2. Rice

Ram *et al.* (1989) studied 60 three-way cross hybrids involving six parents in triallel analysis in rice. The trait 100 seed weight exhibited predominantly epistatic gene action. The magnitude of additive x additive type of epistasis was maximum followed by additive, additive x dominance and dominance x dominance gene effects respectively.

Ram *et al.* (1990) studied 60 three-way crosses involving six parents in rice *viz.*, RP 1570-6-5-1-1, Mashuri, IR 52, TAU 18 and Pawanpeer.

Days to panicle emergence, panicle length and flag leaf area were analysed adopting the triallel analysis model and these characters showed predominant epistatic genetic variances. Among the genetic components, additive x dominance was maximum for days to panicle emergence and flag leaf area and dominance x dominance was maximum for panicle length.

Ram *et al.* (1994) concluded that the superiority of the triplets may be due to (i) the involvement of atleast one parent showing better general combining ability general line effects with the restriction that it should be placed in a proper order or (ii) either cross showing better two line specific effects or rarely the presence of interaction between the three poor general combiners used in particular order.

Parthasarathy (1994) studied 60 three-way cross hybrids involving six parents in a triallel analysis in rice. The genetic components of variances in triallel revealed a reverse trend of the diallel, where in the epistatic genetic action accounted increased estimates over additivity for all the six characters namely days to 50 per cent flowering, productive tillers per plant, panicle length, grains per panicle, 100 grain weight and yield per plant.

6.3.1.3. Wheat

Joshi (1990) carried out triallel analysis in wheat. The spring x winter wheat cross derivatives of PC 1031 and P 823 were crossed with four spring wheat varieties to produce 60 three way crosses. Both PC 1031 and PC 823 were found to be good general combiners for a number of yield contributing traits. Most of the three-way crosses which recorded high yield had invariably either PC 1031 or PC 823 or both as parents. The study indicated the possibility of improving spring wheat by using spring and winter wheat derivative in hybridization programme.

6.3.1.4. Barley

Chaudhary and Singh (1976) studied 60 three-way hybrids involving three in each of exotic and indigenous strains of barley for number of spikes per plant. The digenic interaction components of genetic variances were many times higher than additive and dominance components. The analysis furnished information on order effect, a change in the specific combining ability effect.

6.3.1.5. Sorghum

Hussain Sahib and Reddy (1989) compared single crosses, three-way crosses and double crosses besides parents in sorghum. The yield of all the three groups of crosses were statistically superior to the respective parents. Three-way and double crosses were on par with single crosses, but recorded numerically more yield. The range for yield was less in double crosses followed by three-way and single crosses which suggested that double crosses had more buffering capacity as compared to three-way crosses and single crosses.

6.3.1.6. Pearl millet

Arunachalam and Balarami Reddy (1979) carried out basic studies on triallel crosses in pearl millet. A set of 60 three way crosses in pearl millet, made in a triallel design with six parents, when evaluated for yield and its components, revealed higher order epistatic interactions for several of them. Additive epistasis of the type additive x additive, additive was high for yield, plant height and days to first flowering. HxL single crosses between parents of high and low overall general combining ability provided potential seed parents of three-way crosses. Populations generated by three-way crosses alone and in combination were found productive. Not only the parents, but the order in which they occur in three-way crosses was important to record good yield improvement.

6.3.1.7. Groundnut

Arunachalam *et al.* (1985) made a set of three-way crosses in a line x tester design in groundnut. The variance of general combining ability was higher than that of specific combining ability for yield components while the situation was reverse for early stage characters.

6.3.1.8. Sesame

Backiyarani (1995) studied diallel and triallel analysis in sesamum involving six parents to estimate the combining ability effects of parents and hybrids, gene action and order of parents for eleven characters. Triallel analysis showed that all the six parents were good general combiners to be used as any one of the grand parents. Both TNAU 65 and CO 1 were found to be useful as immediate parents for improving yield and physiological traits. The performance of three-way cross hybrids was very much influenced by the parents with variable general and two line specific effects. There was a direct relationship between the highly significant three line specific effect and high mean performance in eight hybrids and this indicated that these hybrids would be useful in heterosis breeding. Order effects in three-way crosses were found important for all the traits.

6.3.1.9. Cotton

Shroff *et al.* (1983) have showed the feasibility of three-way cross hybrids in American cotton based on cytoplasmic-genetic male sterile line as follows:

A x NR (Non-restorer)
↓
CMS F_1 x R (Restorer)
↓
CMFHY (Commercial Fertile hybrid)

Thereby he produced 48 three-way cross hybrids in cotton for the first time and evaluated for their stability. He suggested that three-way hybrids were found to be more stable on an average. Further Nizama *et al.* (1988) have also developed 21 three-way hybrids by employing cytoplasmic-genetic male sterile lines.

Shroff *et al.* (1989) did comparison of single cross hybrids with three-way hybrids and reported that populational buffering may have an edge over individual buffering in importing stability of performance as the reason for superiority of three-way hybrids in cotton.

Ansingkar *et al.* (1992) studied the feasibility of three-way crosses for exploitation of heterosis in Asiatic cotton. He reported that significant heterosis for seed cotton yield per plant and ginning out turn. He also reported that no heterosis or very negligible heterosis over mid-parent for halo length.

6.3.2. Methods

A three-way cross symbolised by A/B/C has been defined as a cross between the line C and the unrelated F_1 hybrid A/B. Lines A and B being designated as grand parental or half parental lines and C known as the full parental line (Rawlings and Cockerham, 1962). Thus for a given set of 'V' lines, the possible number of three-way crosses would be [V(V-1) (V-2)]/2. In the present study the varieties being six, 60 three-way cross hybrids are possible. The expectation of mean square for the cumulative effect of triallel cross G(ij)k assumes the following model (Ponnuswamy *et al.*, 1974).

$Y_{ijkl} = m + b_1 + h_i + h_j + d_{ij} + g_k + s_{ik} + s_{jk} + t_{ijk} + e_{ijkl}$

where

Y_{ijkl} = Phenotypic value in the $_1^{th}$ replication on $_{ij}^{th}$ cross (grand parents) mated to $_k^{th}$ parent

m = General mean

b_1 = Effects of $_1^{th}$ replication

where

i and j are grand parents and k is the parent

g_{ij} = Average effects of F_1 hybrids

h_i = General line effect of $_i^{th}$ parent as grand parent (first kind general line effect)

h_j = General line effect of $_j^{th}$ parent as grand parent (first kind general line effect)

d_{ij} = two line (i x j) specific effect of first kind (grand parents)

j_k = general line effect of k as parent (second kind effect)

s_{ik} = two line specific effect where, i is the half parent and k is the parent.

Hence specific effect of second kind

t_{ijk} = Three line specific effect

e_{ijkl} = Error effect

The order effect is estimated as the difference between h_i and $g_{i,}$ s_{ij} and s_{ji} and between d_{ij} and $(s_{ij}) + (s_{ji})/2$. The order effects by the parental line $(_k)$ combinations with reference to grand parental lines $(_i)$ and grand parental combinations $(_{ij})$ is also estimated. The presence of order effect does not effect the estimates of the parameters.

Table 56 : The ANOVA for triallel analysis thus,

Source	**df**	**Expected mean squares**
Replication	(r-1)	
Due to g ignoring h	v-1	σ^2
Due to h eliminating g	v-1	$\sigma^2 e+[rv(v-2)(v-3)/(v-1)]\Sigma hi^2$
Due to d eliminating s	v(v-3)/2	σ^2_e

Due to s eliminating d	v^2-3 v+1	σ^2e+[r/(v^2-3v+1)]ΣΣ sij [(v^2-5v+5) sij - sji]
Due to t	v(v^2-6v+7)/2	σ^2e+[2r/v(v2-6v+7)]ΣΣΣtijk^2
Due to crosses	p-1	σ^2e+[2r/v(v-1) (v-2)-2] ΣΣΣC2ijk
Due to g eliminating h	v-1	σ^2e+[rv(v-3)/(v-1)]Σg^2i
Due to d eliminating s	v(v-3)/2	σ^2e+[2(v-1)(v-4)/v(v-3)^2] ΣΣdij
Error	(r-1)(p-1)	σ^2e

(Source: Thirugnanakumar, 1991)

where,

'v' stands for number of parents and p = v (v-1) (v-2)/2.

All the effects are to be tested against error mean squares using appropriate degrees of freedom. The various effects under traillel analysis are,

I. h_i = General line effect of first kind (grand parent)

This is infact the general combining ability effect of a line used as one of the grand parents

$$hi = \frac{(v-1)}{rv(v-2)(v-3)}[Yi.. + [(v-4)/(v-1)]Y..i. - [(v-4)/(v-1)]Y...]$$

II. g_k = General line effect of the second kind. This refers to the general combining ability of a line used as parent crossed to the single hybrid.

$$G_k = \frac{v-4}{v(v-3)r}[Y..i. + [1/v-2]\,Yi...-[1/v-2]\,Y....]$$

III. d_{ij} = Two line specific effect of first kind (grand parenst):

$$d_{ij} = \frac{v-3}{(v-1)(v-4)r}\left(Yij.. + [1/(v-3)]Yi.j. + Yj.i.) - [2/v(v-3)]\right)$$

$$Y... - [(v^2 - 4v + 2)r/(v-3)](h_i + h_j) - (r/(v-3)(g_i + g_j))$$

IV. s_{ik} = Two line specific effect where i is half parent and k is parent i.e., specific effect of second kind.

s_{ik} = (D/D_2)[Yi.k.+(1/D)Yj.i.+(v-3)/D)Yij..-(2(v-3)/vD)Y..-r(v-2) hj-((v-2)D)rhj-(r/D)g_k -(D_1r/D)g_k]

where,

D $= V^2 - 5v + 5$

$D_1 = V^3 - 7v^2 + 4\ v - 7$

$D_2 = r(v-1)\ (v-3)\ (v-4)$

V. t_{ijk} = Three line specific effect.

$= [\overline{Y}ijk\cdot - \overline{Y}.... - hi - hj - gk - dij - sik - sjk]$

The standard errors to test their effects are computed as follows.

S.E. $(h_i) = [(v-1)\ \sigma^2_e / rv^2\ (v-2)\ (v-3)]^{1/2}$

S.E $(g_i) = [2(v-1)\ \sigma^2_e / rv^2\ (v-3)]^{1/2}$

S.E. $(s_{ij}) = [(v-2)^2\ (v^2-4v+1)\ \sigma^2_e / (v-1)^2 rv(v-1)\ (v-4)]^{1/2}$

S.E. $(d_{ij}) = [(v-3)^2\ \sigma^2_e / r(v-1)^2\ (v-4)]^{1/2}$

S.E. $(t_{ijk}) = [(v^2-6v+7)\ \sigma^2_e / r\ (v-1)(v-2)]^{1/2}$

The level of significance of the various effects was tested at p = 0.05% as, CD + SE x t = variance x t (1.96).

Estimation of parameters useful for the evaluation of lines

i) $s^2tij = [1/(v-3)]\ Stijk^2 - (v^2 - 6v + 7)\ s^2/r(v-1)\ (v-3)$

ii) $s^2ti.j = [1/(v-3)]\ Stijk^2 - (v^2 - 6v + 7)\ s^2/r(v-1)\ (v-3)$

iii) $s^2tji. = [1/(v-3)]\ Stikj^2 - (v^2 - 6v + 7)\ s^2/r(v-1)\ (v-3)$

iv) $s^2ti.. = [1/(v-2)\ (v-3)]\ SStijk^2 - (v^2 - 6v + 7)\ s^2/r(v-2)\ (v-3)$

v) $s^2t..i = [2/(v-2)\ (v-3)]\ SStjki^2 - (v^2 - 6v + 7)\ s^2/r(v-2)\ (v-3)$

vi) $s^2di. = [1/(v-2)]Sdij^2 - (v-3)^2\ s^2/r(v-1)(v-2)\ (v-4)$

vii) $s^2si. = [1/(v-2)]\ (Ssij^2 - (v-2)\ (v^2-4v+1)\ s^2/r(v-1)(v-3)$

viii) $s^2s.i = [1/(v-2)]\ (Ssji^2 - (v-2)\ (v^2-4v+1)\ ^2/r(v-1)(v-3)$

ix) $s^2gi = (gi^2)-2\ (v-1)\ s^2/rv^2(v-3)$

x) $s^2hi = hi^2-\ (v-1)^2\ s^2/rv^2(v-2)\ (v-3)$

These statistics are required to assess the relative importance of general and specific effects. For instance, if the variances σ^2di., σ^2si. and σ^2ti... are substantially larger than σ^2s i, σ^2t.. i in that case to get a high yielding combination, there is need to use line i as one of the grand parents rather than using it as a parent. Further, the relative importance of s^2h_i and s^2g_i as compared to s^2 si·, s^2 di· and s^2 ti.. will provide idea whether the general line effects alone are important or two line and three line specific effects are also to be considered for evaluating the breeding potential of lines (Ponnuswamy *et al.*, 1974).

The genetic components of variances were obtained as follows assuming F = 1 using the intermediary components of T, DS, GH, M (t), M(E), M(g/h), M(d/s), M(s/d) and M(h/g) as,

Additive : $\sigma^2_A = (1/227F)\ [(448s^2_h + 40s^2_g + 604s_{gh} - 292s^2_d - 584s_{ds})]$

Dominance :

$\sigma^2_D =$

$$\left(\frac{1}{127F^2}\right)\left[416\sigma^2_h - 352\sigma^2_g + 496\sigma_{gh} - 336\sigma^2_d - 672\sigma_{ds} - \frac{1816}{3}\sigma^2_s + \frac{4540}{3}\sigma_{ss} - 254\sigma^2_t \frac{3556}{3}\sigma_{tt}\right]$$

Additive

$$\textbf{x}: \sigma^2_{AA} = [\frac{1}{227F^2}][(-832\sigma^2_h + 704\sigma^2_g - 992\sigma_{gh} + 672\sigma^2_d + 1344\sigma_{ds})]$$

Additive

Additive

$$\textbf{x}: \sigma^2_{AD} = [(\frac{32}{3F^3})][\sigma^2_s - \sigma_{ss} + 4\sigma_{tt}]$$

Dominance

Dominance

$$\textbf{x}: \sigma^2_{DD} = [(\frac{1}{3F^4}) - 16\sigma^2_s + 16\sigma_{ss} + 24\sigma^2_t - 32\sigma_{tt})]$$

Dominance

where the intermediary components are,

σ^2_t = $\{[v^2\text{-}6v+6)\ (v^2\text{-}6v+7)/r(v\text{-}1)\ (v\text{-}2)\ (v\text{-}4)\ (v\text{-}5)]M(t) + [2rT/v(v^2\text{-}6v+7)\ (v^2\text{-}6v+6] - [1\text{-}2r/(v^2\text{-}6v+7)\ (v^2\text{-}6v+6)]\ M(E)\}$

σ_{tt} = $[(v^2\text{-}6v+7)/(v\text{-}1)\ (v\text{-}2)\ (v\text{-}4)\ (v\text{-}5)]\ [(1/v)T + (1/r)M(t) + \{(1\text{-}(1/r)\}\ M(E)]$

σ_d = $[(v\text{-}3)/r(v\text{-}1)\ (v\text{-}4)]\ [\{M(d/s) - ((2r/v\text{-}3)\ \sigma_{tt}\text{-}r\ \sigma^2_t - M(E)]$

σ_{ds} = $[(1/v(v\text{-}3)]\ [DS - (V(v\text{-}2)(v\text{-}3)(v\text{-}1)(v\text{-}4))\sigma_{tt} - (v(v\text{-}3)/(v\text{-}1)(v\text{-}4)\ \sigma^2_t + M(E)]$

σ^2_s = [(v-2) (v²-3v+1)²/r(v-3)(v-1)² (v²-6v²+8v+1)] (M(s/d) + 4r (v-1) (v-3) SS / (v-2) (v²-3v+1)²] – (r(3v⁴-21v³+41v²-32v +10) σ_{tt}/ (v-2) (v-4) (v²- 3v+1)²] - rσ^2_t(1+4/(V-4) (v²-3v+1)²)- [{(1+4r/(v-4) (v²-3v+1)²M(E)]

σ_{ss} = [2(v²-4v+1)/v-1) (v³-6v²) +8v +1)] [(s.s) + (v-2) (v²-3v +1)M(s/d)/2r (v-1) (v-3) (v²-4v+1) - (v⁶-11v⁵+44v⁴-77v³+55v²-9v-4)] $_{tt}$/2(v-1) (v-3) (v-4) (v²-4v +1)] – [(v-2) (2r(v² -4v +2) + (v-4) (v²-3v+1)) M(E)/2r((v-1) (v-3) (v-4) (v²4v +1)]

σ_{gh} = [1/(v-1)] [(GH-(2(v-1)- (σ_{ds} + σ^2s)/(v-2))] – [2 (v-1)²]

$$\sigma_{ss}/v(v-2)-\frac{2(v-1)^2}{v(v-3)}\sigma_{tt}\frac{2(v-1)}{(v(v-2)(v-3)}\{\sigma^2_t+M(E)\}]$$

$$\sigma^2_h = [\frac{(v-1)}{rv(v-2)(v-3)}][M(h/g)-\frac{v(v-3)}{(v-1)}\sigma^2_d\frac{2r(v-3)v}{(v-1)}-$$

$$\sigma ds(v-3)r\sigma^2_s+[\frac{r(v+1)}{(v-1)}](\sigma_{tt}-r\sigma^2_t]-M(E)$$

$$\sigma^2_g = [\frac{2}{r(v-3)v}][M(g/h)-[\frac{2r(v-1)(v-3)}{(v-2)}\sigma^2_s[\frac{2r(v-3)}{(v-2)}]\sigma_{ss}-$$

$$\frac{2r}{(v-2)}\sigma_{tt}-r(\sigma^2_t)-M(E)$$

$\sigma^2 e$ = error mean square

T = $\Sigma\Sigma\Sigma\ t_{ijk}\ (t_{ijk} + t_{jki})$

DS = $\Sigma\Sigma\ d_{ij}\ (s_{ij} + s_{ji})/2$

SS = $\Sigma\Sigma\ s_{ij}\ s_{ji}$

GH = $\Sigma\ h_i g_i$

M(h/g) = Mean sum of squares h_i adjusted for g_i

M(s/d) = Mean sum of squares s_{ij} adjusted for d_{ij}

M(t) = Mean sum of squares t_{ijk}

M(E) = Mean sum of squares of error

M (g/h) = Mean sum of squares g_i adjusted for h_i

M (d/s) = Mean sum of squares d_{ij} adjusted for s_{ij}

6.3.3. Example

Eswaran (2007) crossed six genotypes of Okra in diallel fashion and produced 15 direct and 15 reciprocal cross hybrids. He kept the 15 direct cross hybrids as seed parents and crossed with four unrelated parents to effect as many as 60 three way cross hybrids. He recorded data on fruit yield and its component characters.

6.3.3.1. Analysis of variance (ANOVA)

The analysis of variance indicated that the crosses differed among themselves for all the five fruit yield and its component characters studied. It indicated the presence of sufficient amount of genetic variability. Therefore, further analysis was appropriate (Table 57).

Table 57 : Triallel-ANOVA for fruit yield and its component characters in Okra

S. No.	Characters	Mean squares		
		Replication	Crosses	Error
1.	Number of branches per plant	0.05	2.51**	0.07
2.	Plant height	52.43	368.88**	7.01
3.	Number of fruits per plant	4.75	90.71**	1.82
4.	Fruit weight	0.84	9.87**	0.35
5.	Fruit yield per plant	132.01	37239.0**	2132.04

** Significant at 1per cent level (Source: Eswaran, 2007)

6.3.3.2. per se performance of three-way cross hybrids

Out of the 60 three way cross hybrids produced by Eswaran (2007), 18 were high yielders. They recorded higher fruit yield per plant than the grand mean of the 60 three way cross hybrids. The mean fruit yield per plant was higher with the three way cross hybrid *viz.,* (Parbani Kranti × Pusa Sawani) × IC 128076. All the three parents of the above said crosses were poor yielder than the grand mean of the parents as well as their single cross hybrids.

When highly performing F_1 hybrids (Pusa Sawani × EC 305626) were crossed with the low performing parents *viz.,* EC 112112 the resulting three way cross hybrids had higher number of branches per plant. On the otherhand, when low performing F_1 hybrids (EC 112112 × EC 305626) were crossed with the high performing parent (Arka Anamika), the resulting three way cross hybrids also had higher number of branches per plant. On the contrary, when the low fruit yielding F_1 hybrids (Parbani Kranti × Pusa Sawani); (Arka Anamika × IC 129076) and (Arka Anamika × EC 305626) were crossed with low performing parents *viz.*, IC-128076; EC 112112 and EC 305626), the resulting three way cross hybrids recorded higher fruit yield per plant (Table 58). The result of Eswaran (2007) indicated that no generalization can be done with regard to the performance of the three way cross hybrids from that of the performance of their parents. It seems that the genetic constitution of the three-way cross hybrids decide the performance of the three way cross hybrids rather than the phenotypic performance of their parents, whether F_1's or lines. It may indicate that the combination of genes decide the fruit yielding potential rather than the phenotypic performance of the parents (heterozygotes or homozygotes).

Table 58 : Mean values of three-way cross hybrids

S. No.	Three-way cross hybrid	Number of branches per plant	Plant height	Number of fruits per plant	Fruit weight	Yield per plant
1.	AA x PK x PS	4.00	155.36	28.00	19.43	543.73
2.	AA x PK x EC- 112112	4.16	160.63	28.70	18.57	528.23
3.	AaxPKxEC-305626	4.26	150.30	29.30	17.13	501.97
4.	AA x PK x IC-128076	5.90**	172.27**	40.70**	21.47	739.90
5.	AA x PS x PK	4.73	147.53	29.47	19.06	559.87
6.	AAxPSxEC-112112	3.80	145.90	37.50**	19.93	747.47
7.	AAxPSxEC-305626	5.10**	136.60	34.70	20.06	696.37
8.	AAxPSxIC-128076	5.30**	139.27	26.27	21.20	446.60
9.	AA x EC-112112xPK	5.97**	146.86	29.50	21.17	624.27
10.	AA x EC-112112 x PS	4.63	135.10	34.76	21.77	756.36
11.	AA x EC-112112 x EC-305626	4.20	145.13	32.63	19.83	648.67
12.	AA x EC-112112 x IC-128076	3.80	163.97**	37.13**	18.00	668.43
13.	AAxEC-305626 x PK	3.50	154.27	34.16	17.20	587.27
14.	AAxEC-305626xPS	3.20	140.67	27.93	18.03	615.70

Contd...

15.	AA x EC-305626 x EC-112112	5.00**	165.53**	43.77**	19.90	555.87
16.	AA x EC-305626 x IC-128076	2.13	140.73	27.40	19.97	873.66**
17.	AA x IC-128076 x PK	2.80	158.03	31.80	20.30	555.76
18.	AA x IC-128076 x PS	3.06	161.50**	35.10**	21.20	674.07**
19.	AA x IC-128076 x EC-112112	4.93**	169.70**	42.33**	21.73	919.93**
20.	AA x IC-128076 x EC-305626	4.96**	154.00	36.40**	20.51	745.93
21.	PK x PS x AA	2.53	145.13	26.40	22.56**	595.43
22.	PK x PS x EC-112112	3.10	151.96	26.73	22.50**	601.33
23.	PK x PS x EC-305626	3.63	160.53**	26.63	23.36**	622.40
24.	PK x PS x IC-128076	4.43	168.87**	44.86**	20.53	921.10**
25.	PK x EC-112112 x AA	4.16	144.53	36.87**	22.60**	777.83**
26.	PKxEC-112112xPS	3.00	163.16**	33.10	22.93**	757.66
27.	PK x EC-112112 x EC-305626	3.60	155.46	37.00**	22.23	822.40**
28.	PK x EC-112112 x EC-305626	3.83	151.67	34.87	19.86	700.87
29.	PKx EC-305626 xAA	4.40	139.60	29.13	22.46**	654.56
30.	PKxEC-305626xPS	3.53	151.47	33.23	20.86	692.60
31.	PK x EC-305626 x EC-112112	5.03**	144.90	20.40	24.36**	496.83
32.	PK x EC-305626 x IC-128076	4.10	162.57**	29.23	22.07	645.06
33.	PK x IC-128076 x AA	3.97	135.47	35.26**	23.07**	775.37**
34.	PKxIC-128076xPS	3.46	144.56	22.83	23.46**	538.53
35.	PK x IC-128026 x EC-112112	2.93	163.00**	26.66	23.57**	625.23
36.	PK x IC-128076 x EC-305626	5.13**	172.63**	26.33	23.56**	621.50
37.	PS x EC-112112xAA	4.96**	169.93**	34.43	28.50**	705.63

Contd...

38.	PS x EC-112112 x PK	4.23	171.10**	22.93	25.90**	593.87
39.	PS xEC-112112 x EC-305626	4.56	168.10**	32.80	21.76	714.43
40.	PS x EC-112112 x IC-128076	4.00	139.13	37.83**	19.66	743.80
41.	PS x EC-305626 x AA	4.06	163.43**	37.73**	22.13	835.60**
42.	PSxEC-305626 xPK	5.73**	152.83	27.80	23.20**	624.36
43.	PS x EC-305626 x EC-112112	6.06**	163.60**	29.60	22.20	657.10
44.	PS x EC-305626 x IC-128076	6.06**	143.30	32.63	22.30	727.30
45.	PS x IC-128076 x AA	5.23**	158.86**	39.06**	20.30	797.23**
46.	PS x IC-128076 x PK	5.20**	140.33	39.40**	21.70	854.83**
47.	PS x IC-128076 x EC-112112	5.36**	156.97	29.60	24.33**	719.70
48.	PS x IC-128076 x EC-305626	4.40	169.20**	36.06**	25.53**	848.80**
49.	EC-112112 x EC-305626 x AA	6.03**	153.20	38.20**	20.80	793.73**
50.	EC-112112 x EC-305626 x PK	5.43**	162.13**	33.46	21.66	722.83
51.	EC-112112 x EC-305626 x PS	4.87**	143.43	37.06**	20.00	740.86
52.	EC-112112 x EC-305626 x IC-128076	4.80**	149.43	36.37**	21.13	768.50**
53.	EC-112112 x IC-128076 x AA	3.53	150.80	38.70**	21.40	828.33**
54.	EC-112112 x IC-128076 x PK	3.80	141.53	34.26	23.33**	810.53**
55.	EC-112112 x IC-128076 x PS	4.03	164.53**	39.30**	21.66	851.33**
56.	EC-112112 x IC-128076 x EC-305626	4.00	170.96**	37.60**	21.20	797.16**
57.	EC-305626 x IC-128076 x AA	4.13	170.30**	36.66**	19.73	723.63

Contd...

58.	EC-305626 x IC-128076 x PK	4.80**	167.57**	21.67	24.70**	534.30
59.	EC-305626 x IC-128076 x PS	4.46	168.20**	37.10**	21.13	783.13**
60.	EC-305626 x IC-128076 x EC-112112	4.40	153.90	33.40	21.20	704.10
	Grand mean	4.34	154.86	33.01	21.37	695.33
	S.E	0.15	1.51	0.77	0.34	26.43
	CD at 1 per cent	0.41	3.97	2.02	0.89	69.27

** Significant at 1per cent level (Source: Eswaran, 2007)

6.3.3.3. Gene action and line effects

According to Ponnuswamy *et al.* (1974) the variance and co-variance components of the general effects *viz.*, σh^2, σg^2 and σgh are the functions of additive and additive × additive type of epistasis only. The components σd^2 and sds are also functions of additive × additive type of epistasis only. $\sigma^2 s$ and σss are the only two components which involve dominance component. The components st^2 and stt are the functions of epistatic components other than additive × additive type.

The non-significance of the three line specific effect will provide positive evidence of the absence of allt he epistasis other than additive × additive. The pooled mean square of 'd' effects and ''t effect when tested against error will show the absence of all the types of epistatic gene action. Similarly, the pooled mean square of 's' and 't' effects when tested against error will provide an indirect test for the absence of all the gene action other than additive and additive × additive type of epistasis. Like wise, the non-significance of the pooled mean square 's', 'd' and 't' effects when tested against error will show the absence of all ther types of non-additive gene action.

The analysis of variance for the three way crosses showed (Table 59) that the general line effect of the first kind (hi) was significant for all the five characters studied by Eswaran (2007). Similarly, the general line effect of the second kind (gi), two line specific effect of both first kind (dij) and second kind (sij) and three line specific effect (tijk) were significant for all the five traits studied by Eswaran (2007). The result indicated the importance of both additive and non-additive gene effects in the expression of the characters studied. The average lines in hybrid combinations, 'hi' and 'gi' indicate the additive and additive type of epistatic gene effects (Rawlings and Cockerham, 1962).

Table 59 : Triallel-ANOVA for combining ability analysis in Okra

S. No.	Source	df	Number of branches per plant	Plant height	Number of fruits per plant	Fruit weight	Fruit yield per plant
1.	General line effect of the first kind (h_i)	5	2.24**	180.21**	11.26**	6.71**	45765.14**
2.	General line effect of the second kind (g_i)	5	1.36**	214.22**	8.53**	7.99**	53736.63**
3.	Two line specific effect of the first kind (d_{ij})	9	5.12**	472.50**	43.43**	5.01**	30518.02**
4.	Two line specific effect of the second kind (s_{ij})	19	1.80**	473.91**	18.97**	7.49**	17714.19**
5.	Third line specific effect (t_{ijk})	21	1.80**	277.14**	22.94**	4.96**	36594.79**
6.	Due to crosses	59	2.51**	368.88**	24.02**	9.87**	37239.07**
7.	Due to hi eliminating g_i effect	5	3.70**	357.27**	25.55**	51.44**	117711.42**
8.	Due to d_{ij} eliminating s_{ij} effect	9	5.65**	379.46**	39.95**	5.75**	23547.90**
	Error	118	0.07	7.01	0.75	0.35	2132.04

** Significant at 1 per cent level (Source: Eswaran, 2007)

The line effects are discussed for fruit yield per plant, reported by Eswaran (2007). The parent *viz.,* IC 128076 recorded positive and significant general line effects of both first kind (hi) and second kind (gi). It indicated this parent was a good general combiner when used as a parent in three way cross-programme. The 'hi' values were higher 'gi' value. The relatively high variations in the specific effects (σ^2di, σ^2si and σ^2s-i) associated with the parents IC 128076, indicated that this line may have showed differences in the transmission of character to all its three way cross hybrids.

The two line specific effects of first kind (dij) were positive and significant in six crosses. The highest dij effect was observed in the cross Parbani Kranti × EC 112112 followed by Pusa Sawani × IC 128076 and Arka Anamika × EC 305626. It indicated their superiority as grand parents in three way crosses. Two line specific effect of second kind (sij) were positive and significant in four crosses. The highest sij effect was recorded by the cross Arka Anamika × EC 112112 followed by Parbani Kranti × IC 128076 (Table 60).

Table 60 : Triallel - Estimates of general line and two line specific effects for fruit yield per plant

S. No.	Parents	General line effect		Two line specific effect of the first kind (d_{ij})					
		First kind (h_i)	Second kind (g_i)	AA	PK	PS	EC-112112	EC-305626	IC-128076
1.	AA	- 48.81**	33.87**	-	-19.84* (-31.42)*	-48.93** (-32.89)**	-2.89 (80.33)**	37.73** (-8.25)	33.94** (-7.75)
2.	PK	- 65.15**	74.60**	-19.84* (4.39)	-	35.94** (-7.36)	53.74** (-51.07)*	-27.91** (-22.02)**	-41.93** (76.07)**
3.	PS	7.42	3.03	-48.93* (-34.32)	35.94** (32.84)	-	-63.06** (-4.60)	37.03** (31.62)**	39.02** (-25.55)**
4.	EC-112112	52.90**	-18.59**	-2.89 (-7.24)	53.74** (32.82)	-63.06** (32.79)	-	-1.79 (6.75)	14.01 (-65.12)**
5.	EC-305626	- 9.93**	2.65	37.73** (36.47)	-27.91** (-25.80)	37.03** (37.69)	-1.79 (-70.71)	-	-45.05** (22.35)**
6.	IC-128076	63.58**	53.62**	33.94** (-0.70)	-41.93** (-8.43)	39.02** (-30.23)	14.01 (46.05)	-45.05** (-8.09)	-
		SE$(h)_i$= 4.53		SE(g_i) = 5.73		SE(d_{ij}) = 7.99		SE(s_{ij}) = 7.02	

Figures in brackets correspond to estimates of s_{ij} (upper half) and s_{ji} (lower half) (Source: Eswaran, 2007)

* Significant at 5 per cent level; ** Significant at 1 per cent level

The three line specific effects (tijk) were found to be positive and significant in 20 three way crosses, among which the highest tijk effect was recorded by (Parbani Kranti × Pusa Sawani) × IC 128076 followed by (Arka Anamika × EC 305626) × IC 128076 and (Pusa Sawani × IC 128076) × Parbani Kranti. In the best performing triplet of (Parbani Kranti × Pusa Sawani) × IC 128076, the lines Parbani Kranti × Pusa Sawani were poor in their grand parental performance. Similarly, the line IC 128076 was also poor in the parental performance. The two line specific effect of first kind (dij) of the cross Parbani Kranti × Pusa Sawani was good (Table 61).

The parent order effect was clearly elucidated in the triplet (Parbani Kranti × Pusa Sawani) × IC 128076 which had the highest positive significant tijk effect. The two other alternative combinations with the same parents namely (Pusa Sawani × IC 120876) × Parbani Kranti and (Parbani Kranti × IC 128076) × Pusa Sawani has positive and negative significant tijk effects, respectively but with lower magnitude.

The components of genetic variance indicated the presence of enormous amount of dominance × dominance and additive × dominance type of epistatic interactions. This variability could be exploited by resorting to hybrid breeding programme.

Table 61 : Triallel-estimates of three line specific effects for fruit yield per plant

S. No.	Grand parental line	Parental line					
		AA	PK	PS	EC-112112	EC-305626	IC-128076
1.	AA x PK	-	-	19.43	-43.95**	-31.92**	56.43**
2.	AA x PS	-	28.05**	-	85.32**	65.34**	-178.17**
3.	AA x EC-112112	-	0.95	56.89**	-	-49.00**	-8.84
4.	AA x EC-305626	-	44.79**	-66.45**	-109.46**	-	131.12**
5.	AA x IC-128076	-	-73.80**	-9.88	68.09**	15.59	-
6.	PK x PS	-82.05**	-	-	2.04	-63.40**	143.41**
7.	PK x EC-112112	9.98	-	-7.62	-	98.18**	-100.54**
8.	PK x EC-305626	-12.49	-	66.91**	44.89**	-	-99.31**
9.	PK x IC-128076	84.57**	-	-78.72**	-2.98	-2.86	-
10.	PS x EC-112112	20.72	-89.79**	-	-	-19.19	88.25**
11.	PS x EC-305626	69.71**	-37.92**	-	21.15	-	-52.94**
12.	PS x IC-128076	-8.38	99.66**	-	-108.52**	17.25	-
13.	EC-112112 x EC-305626	-5.87	53.91**	-69.16**	-	-	21.13
14.	EC-112112 x IC-128076	-24.83*	34.91**	19.89	-	-29.98**	-
15.	EC-305626 x IC-128076	-51.34**	-60.78**	68.70**	43.42**	-	-

SE (t_{ijk}) – 10.95 (Source: Eswaran, 2007)

** - Significant at 1 per cent level * - Significant at 5 per cent level

6.4. Quadriallel analysis

Different techniques have been evolved to estimate the genetic potential of genotypes to be used in the hybridization programme to enhance the yielding potentiality of the crop plants. Rawlings and

Cockerham (1962) have proposed the use of double cross populations from which informations on gene actions and clarification on the underlying genetic systems are obtained which provide the method for inferring the relative importance of different kinds of gene action. It is actually a diallel cross among F_1's which in turn are diallel among parents with the restriction that no parent can appear twice in the same double cross. This analysis will give informations on the order effects of double cross hybrids.

6.4.1. Methods of producing double cross hybrids

A number of inbreds say, six inbreds will be crossed in diallel fashion. The resulting 15 direct single cross hybrids will be mated in diallel fashion with the restriction that only unrelated crosses should be involved in the crossing programme, then say, 45 double crosses would be possible.

6.4.2. Statistical analysis

Rawlings and Cockerham (1962) assigned the statistical model for double cross hybrids for estimates of genetic components of epistatic nature and information on order effects of combinations. Singh and Chaudhary (1977a, b) presented an elaborate work on the linear model of the double cross hybrids. The model is as follows.

$y(ij)\,(kl)_m = m\,rm + G(ij)\,(kl) + e(ij)\,(kl)_m$

where

$y(ij)\,(kl)_m$ is the observation on double cross

(ij) (kl) grown in replications

m, m - 1, 2, ..., r, i, j, k, l = 1, 2, ... P

where no two of i, j, k, and l can be the same

m = general mean,

rm = the effect of the mth replication

G(ij) (kl) = the genotypic effect of the double cross hybrid (ij) (kl)

e(ij) (kl) = a random error

Further,

G(ij), (kl) = (gi + gj + gk + gl) + (Sij + Sik + Sil + Sjk + Sjl + Skl) + (Sijk + Sijl + Sikl + Sjkl) + (Sijkl) + (tij + tkl +tlk + tj.l + tj.k + tj.l) + (tij.l + tij.k + tki.j + tkl.j) + (tij.kl)

where,

gi = the average general effect of line i.

Sij = the 2-line interaction effect of lines i and j appearing together, irrespective of arrangement

Sijk = the 3-line interaction effect of lines i, j and k appearing together, irrespective of arrangement

Sijkl = the 4-line interaction effect of lines i, j, k and l appearing together, irrespective of arrangement

tij = the 2-line interaction effect of lines i and j due to the particular arrangement (ij) (- -)

ti.j = the 2-line interaction of lines i and j due to the particular arrangement (i-) (j-)

tij.k = the 3-line interaction effect of lines ij and k due to the particular arrangement (ij) (k-)

tij.kl = the 4-line interaction effect of lines i, j, k and l due to the particular arrangement (ij) (kl).

The procedure for estimating general and specific line effects of various arrangements using the least square technique (Singh and Chaudhary, 1977a) are given below.

i) Average effect of line i = gi = (yi...)/r P_1 P_2 P_3/2) m.

ii) The two-line interaction effects of lines i and j appearing together irrespective of arrangement = Sij = [(yij..)/(3r P_2 P_3/2)] – m – gi – gj.

iii) The three-line interaction effects of lines i, j and k appearing together irrespective of arrangement = Sijk = [(yijk..)/3r P_3] – m gi – gj – gk. – Sij – Sik – Sjk.

iv) The four line interaction effects of lines i, j, k and l appearing together irrespective of arrangement = Sijkl = [Yijkl../(3r)] – m – gi – gj – gk – gl – Sij – Sil – Sik – Sjk – Sjl – Skl – Sijk – Sijl – Sikl – Sjkl.

v) The two line interaction effects of lines i and j due to the particular arrangement (ij) (- -) = tij = [y(ij) (..)/(r P_2 P_3/2)] – m – gi – gj – Sij

vi) The two line interaction effects of line i and j due to the particular arrangement (i-) (j-) = ti.j = (t(i-) (j-) = y(i.) (j.)/r P_2 P_3 – m – gi – gj – Sij

vii) The three line interaction effects of lines i, j and k due to the particular arrangement (ij) (k-) = tijk = t (ij) (k-) = [r(ij) (k-)/r P_3] – – gi – gj – gk – Sij – Sik – Sjk – Sijk – ti.j – tj.k – ti.k

viii) The four line interaction effects of line i, k, j and l due to the particular arrangement
(ij) (kl) = tij.kl = (y(ij) (kl)/r)– – gi – gj – gk – gl Sij – Sik – Sjk – Sil – Sjk – Sjl – Skl – Sijk – Sijl – Sikl – Sjkl – tij – tkl – ti.k – ti.l – tj.k – tj.l – tij.k – tij.l – tkl.i – tkl.j

where,

Pi denotes (P – i) for example, with P – 6, P value are P – 6 – 0 = 6, P_1 = 6 – 1 = 5 and so on.

The ANOVA for the different effects would be then

Source	**df**	**MS**
Replications	(r – 1)	R
Total	3r 6 C_{4-1}	-
Hybrid	3^6 (4 – 1)	H
Error	(r – 1) (3^6 C_{4-1})	E
1 line general	P_1	G
2 line specific	$PP_3/2$	S_2
3 line specific	$PP_1P_5/6$	S_3
4 line specific	$PP_1P_2P_7/24$	S_4
2 line arrangement	$PP_3/2$	T_2
3 line arrangement	$PP_2P_4/3$	T_3
4 line arrangement	$PP_1P_4P_5/12$	T_4

(Source: Singh and Chaudhary, 1977)

The minimum number of lines required for quadriallel is eight. If P = 7, 3 and 4-line specification is treated as specific single sum with 14 df. With P = 6, only one 2 line specific sum of squares is available with 9 df and variances due to 3 and 4 line specific effects cannot be estimated. The negative variances were however, required in further calculation of these effects. With the help of mean sum of squares, estimates of components of variances are calculated as

$S^2\, t_4 = (T_4 - E)/r$

$S^2\, t_3 = T_3 - T_4/r\, P_3$

$S^2\, t_2 = (3/r\, P_1P_2)\, (T_2 - (2\, P_2/P_3)\, T_3 + (P_1/P_3)\, T_4)$

$S^2\, S_4 = (S_4 - E)/3r$

$S^2\, S_3 = (S_3 - S_4)/3r\, P_6$

$S^2\, S_2 = (2/3r\, P_4P_5)\, (S_2 - (2\, P_5/P_6)\, S_3 + (P_4/P_6)\, S_4$

$S^2G = (2/r\, P_2P_3P_4)\, (G - 3\, P_3/P_5)\, S_2 + (3\, P_2/P_6)S_3 - (P_2P_3/P_5P_6)\, S_4)$

where,

$C = \Sigma y^2 \times\times\times\times\times \,/\, r\, PP_1P_2P_3,\; M = y^2(ij)\,(kl)_m^{-C}$

$R = (8\, \Sigma y^2 \times\times\times\times\times\, m) \,/\, (PP_1P_2P_3) - C$

$H = (\Sigma y^2\, (ij)\, (kl)/r) - C$

$G = (2\Sigma y^2 i\, .../r\, P_2P_3P_4) - (4P_1/P_4)C$

$S_2 = (2\Sigma y^2 ij\, ... \,/\, 3r\, P_4P_5) - (6\, P_2P_3/P_4P_5)\, C - (3P_3/P_5)G$

$S_3 = (\Sigma y^2 ijk./3r\, P_6) - (4\, P_3/P_6)\, C - (3P_4/P_6)G - (2\, P_5/P_6)\, S_2$

$S_4 = (\Sigma y^2 ijkl/3r) - C - G - S_2 - S_3$

$T_2 = (2\Sigma y^2(ij)\, (..) \,/\, r\, P_1P_2) + (\Sigma y^2(i.)\, (j.)/r\, P_1\, P_2) - (2\Sigma y^2 ij\, .../3\, r\, P_1P_2)$

$T_3 = (\Sigma y^2(ij)\, (k.)/r\, P_3) - (\Sigma y^2 ijk\, .../3r\, P_3) - (2\, P_2/P_3)T_2$

$T_4 = (\Sigma y^2\, (ij)\, (kl)/r) - (\Sigma y^2 ijkl./3r) - T_2 - T_3$

Using the estimate of components of variance, the genetic components of variances were estimated assuming a restricted genetic model. This has been done keeping in view that only the lower order components are generally of interest (Rawlings and Cockerham, 1962). Thus, a set of six estimates of genetic components of variances of the lowest order were derived using either a simple or weighted average of the two components which have the same genetic composition, S^2_{S4} and S^2_{t4}. If k be the simple average, then $k = S^2_{S4} + S^2_{t4}/2$. The six genetic variance were computed keeping F = 1 as,

$$S^2_{10} = (4/3F)\left(6\, S^2_g - 3\, S^2_{s_2} + 2\, S^2_{s_3} + (4/3)\, S^2_{t_2} - 2\, S^2_{t_3} + 2k\right)$$

$$S^2_{01} = \left(S/F^2\right)\left(2\, S^2_{t_2} - 4\, S^2_{t_3} + 3k\right)$$

$$S^2_{20} = (32/F^2)(S^2_{S_2} - S^2_{S_3} - (4/9)\, S^2_{2} + S^2_{t_2} - k)$$

$$S^2_{11} = (128 / F^3)\ (S^2 t_3 - k)$$

$$S^2_{02} = 128 / F^{4(K)}$$

$$S^2_{30} = (256 / 3\ F^3)\ (S^2_{S_3} - S^2_{t_3} + k)$$

where,

S^2_{10} = Additive genetic variance

S^2_{01} = Variance due to dominance deviation

S^2_{20} = Additive × additive component of variance

S^2_{11} = Additive × dominance component of variance

S^2_{02} = Dominance × dominance component of variance and

S^2_{30} = Additive × additive × additive component of variance.

The effects arising due to the arrangements of lines are exclusively the results of dominance effects or interactions involving dominance components.

6.4.3. Example

Subbalakshmi (1989) crossed six sesame inbreds in diallel fashion. She again crossed the 15 single cross hybrids of diallel origin in diallel fashion to evolve 45 double cross hybrids. The result obtained by Subbalakshmi (1989) are presented here under (seed yield only).

The variance due to mean seed yield was highly significant in all the crosses (Table 62).

Table 62 : ANOVA of double crosses for different characters – Mean sum of squares

Source	df	Plant height (cm)	No. of branches	No. of capsules	Capsule length (cm)	Seed yield (g/pl.)	Oil content (%)
Treatment	44	36.0165**	0.4832**	341.9397**	3.4198**	4.1062**	11.8106**
Error	88	8.5167	0.0931	35.3045	0.0922	0.1803	0.1225

**Significant at 1 per cent level. (Source: Subbalakshmi, 1989)

6.4.5. Double cross hybrid analysis

Variances due to the one line general effects, 2 line specific effects as well as 2 line, 3 line and 4 line arrangement effects were all highly significant (Table 64). The individual general effects the parents were not significant. The specific two line effects of i and j lines due to t_{ij} arrangements were significantly positive for 1.2, 1.5, 2.3, 2.5, 4.5 and 4.6 and negatively for 1.4, 1.6 and 3.5. The arrangement $t_{i.j}$ specific effects were positively significant for 4.1, 5.2 and 6.1 and negatively for 5.4. The S_{ij} arrangements were not significant for any of the combinations (Table 63).

The three line interaction effects for the arrangements $t_{ij.k}$ was significant and positive for the combinations 25.4, 24.5, 26.1 and 56.2 and negatively significant for 46.5, 46.2 and 56.1. The combinations 14.5 and 15.6 showed significantly negative specific three line effects irrespective of the arrangements. The four line effects of any individual combinations were not significant (Table 64 and 65).

6.4.6. Estimates of genetic components

The magnitude of the additive (A) variance was low (2.45) while dominance variance (D) was high (28.59). Among the interaction effects, the highest magnitude was found in Additive × Dominance (AD) in the negative direction. The three factor additive^3 interaction variances was greater in magnitude than the two factor additive interaction variances (Table 66).

6.5. Heterosis

Exploitation of hybrid vigour in the field of agriculture is one of the most promising developments in crop breeding programmes. Shull (1948) explained that heterosis is the genetic expression of the beneficial effects of hybridization. A new term heterobeltiosis was coined by Fonseca and Patterson (1968), refers to the improvement over better parents. However standard heterosis is considered to be more valuable as the hybrids found superior than standard variety recommended for commercial cultivation. Heterosis representing deviation of F_1 plants from the mid-parent value, better parent (of that cross) value and the standard variety has been computed from line × tester and diallel analysis.

Table 63 : Estimates of one and two line general and two line arrangement effects for seed yield (g) in double crosses

	1 line general effect	2 line interaction effect of lines i and j due to the particular arrangement (ij) (- -) i.e. tij (above the diagonal) and (i-) (j-) i.e ti.j below the diagonal values in brackets correspond to Sij i.e. effect of i and j irrespective of arrangement					
		1	2	3	4	5	6
1	-0.271		0.5056* (-0.0648)	0.2861 (0.0185)	-0.6824* (-0.0568)	0.5611* (-0.0631)	-0.6704* (-0.1050)
2	0.164	-0.2528		0.4991* (0.1333)	0.2722 (-0.0226)	0.7713* (0.1382)	0.0389 (-0.0198)
3	0.193	-0.1439	-0.2495		-0.2241 (0.0265)	-0.5907* (0.0998)	0.0296 (-0.0856)
4	-0.116	0.3412-	0.1361	0.1120		0.6889* (-0.0320)	0.4898* (-0.0313)
5	0.288	-0.2806	0.3856*	0.2954	-0.3444*		0.1120 (0.0643)
6	-0.177	0.3352*	-0.0194	-0.0148	-0.2449	-0.0568	

SE (gi) = 0.5779 SE (tij) = 0.1714 SE (ti.j) = 0.1561 SE (sij) = 0.1065 (Source: Subbalakshmi, 1989)

*Significant at 5 per cent level

Table 64 : Estimates of three line interaction effects of lines ij and k due to the particular arrangement (ij) (k) ie., tij.k in double crosses for seed yield (g) values in brackets correspond to Sijk, i.e., 3 line effect irrespective of the arrangement

	1	2	3	4	5	6
12			0.6593 (0.0583)	-0.5755 (-0.0700)	-0.0625 (-0.0230)	-0.5269 (-0.0950)
13		0.218		0.0125 (0.0531)	-0.2657 (0.0122)	-0.0546 (-0.0867)
14		0.2287	-0.0114		0.4296 (-0.6920*)	0.0255 (-0.0098)
15		-0.4958	-0.1338	-0.1523		0.2208 (-0.6235*)
16		0.4981	-0.3810	0.3741	0.1792	
23	-0.6810			0.0213 (0.0568)	-0.1356 (0.1835)	0.2963 (-0.0320)
24	0.3468		-0.2162		0.0614* (-0.0013)	0.1403 (-0.0307)
25	0.5583		-0.3969	6.5660*		0.1697 (0.1181)
26	6.6257*		0.2932	-0.0819	0.1889	
34	-0.0111	0.1949			-0.0231 (-0.0004)	0.0634 (-0.0567)
35	0.3995	0.5324		-0.0509		-0.2903 (0.0043)
36	0.4356	-0.4995		-0.0949	0.1292	
45	-0.2773	-0.5014	0.0741			0.0157 (0.0296)
46	-0.3995	-6.6583*	0.0315		-0.0634*	
56	-6.4000*	6.6792*	0.1611	0.0477		

SE (tkj.k) = 0.4424 SE (Sijk) = 0.2282 *Significant at 5 per cent level.
(Source: Subbalakshmi, 1989)

Table 65 : Estimates of four line effects of lines i, j, k and l due to the particular arrangements (ij) (kl) i.e. tij.kl in double crosses for seed yield (g). Figures in brackets are 4 line effects irrespective of their arrangement i.e. Sijkl

	23	24	25	26	34	35	36	45	46	56
12					-0.3671 (0.1626)	0.6718 (0.1998)	-0.4046 (-0.1874)	-0.3046 (-0.2728)	0.6718 (-0.1007)	-0.2671 (0.0032)
13		0.4037	-0.5505	0.1468				0.1468 (-0.0468)	-0.5505 (0.0437)	0.4037 (-0.1163)
14	-0.1366		0.0176	0.1190		0.1190	0.0176			-0.1366 (0.0426)
15	-0.1213	0.3870		-0.2657	-0.2557		0.3870		-0.1213	
16	0.2579	-0.7967	0.5329		0.5329	-0.7967		0.2579		
23								0.2579 (0.1337)	-0.1213 (0.1257)	-0.1366 (-0.2170)
24						-0.7967	0.3870			0.4037 (0.1343)
25					0.5329		0.0376		-0.5505	
26					-0.2657	0.1190		0.1468		
34										-0.2671 (-0.0880)
35									0.6718	
36								-0.4046		

SE (tij.kl) = 1.2655 SE (Sijkl) = 0.5579 (Source: Subbalakshmi, 1989)

Table 66 : Estimates of genetic components of variance for seed yield (g) in double crosses

Components		Magnitude
Additive	A	2.4515
Dominance	D	28.5854
Additive × Additive	AA	-29.0230
Additive × Dominance	AD	-113.8104
Dominance × Dominance	DD	87.7942
Additive × Additive × Additive	AAA	75.2859

(Source: Subbalakshmi, 1989)

6.5.1. Statistical analysis

6.5.1.1. Computation of heterosis in single cross hybrids

The magnitude of heterosis was calculated as follows

(i) Increase of mean F_1 performance over the mean performance of the mid parent

$$\text{Relative heterosis } (d_i) = \frac{\overline{F_1} - \overline{MP}}{\overline{MP}} \times 100$$

(ii) Increase of mean F_1 performance over that of the mean performance of better parent.

$$\text{Heterobeltiosis } (d_{ii}) = \frac{\overline{F_1} - \overline{BP}}{\overline{BP}} \times 100$$

(iii) Increase of mean F_1 performance over that of mean performance of standard variety

$$\text{Standard heterosis } (d_{iii}) = \frac{\overline{F_1} - \overline{SV}}{\overline{SV}} \times 100$$

Standard variety = Parbhani Kranti

(iv) The significance of heterosis was tested using the formula suggested by Wynne *et al.* (1970).

$$\text{'t' for relative heterosis } (d_i) = \frac{\overline{F_1} - \overline{MP}}{\sqrt{(3/2r)\sigma^2_e}}$$

$$\text{'t' for heterobeltiosis } (d_{ii}) = \frac{\overline{F_1} - \overline{BP}}{\sqrt{(2/r)\sigma^2_e}}$$

't' for standard heterosis (d_{iii}) $= \dfrac{\overline{F_1} - \overline{SV}}{\sqrt{(2/r)\sigma^2_e}}$

where,

r = number of replication

σ^2_e = error mean square obtained from ANOVA

6.5.1.2. Computation of heterosis in three way cross hybrids

The mean values of three-way cross hybrid were used for calculation of heterosis.

Relative heterosis (di)

The superiority of F_1 over the mid-parent was estimated as follows.

$$di = \frac{\overline{F_1} - \overline{MP}}{\overline{MP}} x100$$

where,

$\overline{F_1}$ is the average value of the three-way cross hybrid

$\overline{MP}$ is the average value of the three parents involved

$$MP = \frac{(\overline{P1} + \overline{P2} + \overline{P3})}{3}$$

Heterobeltiosis (dii)

The superiority of over the better parent involved in the particular three-way cross was estimated as follows.

$$dii = \frac{\overline{F_1} - \overline{BP}}{\overline{BP}} x100$$

where,

$\overline{F_1}$ is the average value of the three-way cross hybrid

$\overline{BP}$ is the average value of the better parent in the cross

Test of significance

Significance for relative heterosis (di) and heterobeltiosis (dii) was tested by using the 't' test as follows.

For (di) $\quad t = \dfrac{F_{1ijk} - MP_{ijk}}{\sqrt{4/3(Me/r)}}$

where,

F_{1ijk} = mean of ijkth F_1 cross

MP_{ijk} = mid-parent value of the ijkth cross

Me = error variance

r = number of replication

For (dii) $t = \frac{F_{1ijk} - BP_{ijk}}{\sqrt{2(Me/r)}}$

F_{1ijk} = mean of ijkth F_1 cross

MP_{ijk} = Better parental value of the ijkth cross

Me = error variance

r = number of replications

6.5.1.3. Example

Eswaran (2007) made 6 × 6 full diallel crosses in Okra and evaluated the 30 hybrids along with their six parents for heterotic potential. The data on heterosis for yield per plant is given below (Table 67).

The result indicated that eight direct as well as eight reciprocal crosses portrayed significant positive relative heterosis (di) for fruit yield per plant. The F_1 hybrids of the direct cross combinations namely, Parbani Kranti × EC 305626; Arka Anamika × Parbani Kranti, Parbani Kranti × EC 112112 recorded maximum positive relative heterosis. Similarly, the F_1 hybrids of the reciprocal cross combinations *viz.,* EC 305626 × Parbani Kranti, EC 112112 × Parbani Kranti and Parbani Kranti × Arka Anamika registered maximum positive relative heterosis. Six direct and six reciprocal cross combinations recorded significant positive heterobeltiosis. Among the direct crosses, it was maximum with Parbani Kranti × EC 305626, Parbani Kranti × Ec 112112 and Arka Anamika × Parbani Kranti. Among the reciprocal crosses it was maximum with Parbani Kranti EC 305626, Arka Anamika × Parbani Kranti and Parbani Kranti × EC 112112. Commercial heterosis (standard heterosis) for this character was significant and positive in five direct and four reciprocal crosses. The F_1 hybrids of the direct cross combinations *viz.,* Arka Anamika × Parbani Kranti, Parbani Kranti × EC 112112 and Parbani Kranti × EC 305626 recorded maximum commercial heterosis. Similarly, the F_1 hybrids of the reciprocal cross combinations namely, Parbani Kranti × Arka Anamika,

EC 305626 × Parbani Kranti and EC 112112 × Parbani Kranti registered maximum significant positive standard heterosis. The result indicated that there exists good scope for fruit yield improvement in Okra through heterosis breeding.

Table 67 : Estimates of heterosis diallel hybrids for fruit yield per plant in Okra

S. No.	Cross combinations	Relative heterosis (d_i)		Heterobeltiosis (d_{ii})		Standard heterosis (d_{iii})	
		Direct	Reciprocal	Direct	Reciprocal	Direct	Reciprocal
1.	AA x PK	47.03**	59.22**	39.76**	51.35**	55.11**	67.97**
2.	AA x PS	34.28**	2.34**	14.87	-12.45	27.49**	-2.84**
3.	AA x EC-112112	-19.73**	-11.59**	-26.91**	-19.50**	-18.88**	-10.66**
4.	AA x EC-305626	-11.10**	-4.94**	-23.11**	-17.79**	-14.66**	-8.76**
5.	AA x IC-128076	-2.77	-4.05	-20.24**	-21.28**	-11.48**	-12.64**
6.	PK x PS	19.47**	-24.38**	-27.97**	-32.36**	-27.97**	-32.36**
7.	PK x EC-112112	46.34**	59.31**	39.84**	52.24**	39.84**	52.24**
8.	PK x EC-305626	62.21**	70.78**	46.80**	54.55**	46.80**	54.55**
9.	PK x IC-128076	18.17**	18.91**	1.11	1.73	1.11	1.73
10.	PS x EC-112112	16.78**	18.01**	8.95**	10.10**	-0.73	0.32
11.	PS x EC-305626	41.34**	4.54**	39.51**	3.18**	12.99	-16.43
12.	PS x IC-128076	-0.80	-3.77	-5.69*	-8.51*	-25.60*	-27.82**
13.	EC-112112 x EC-305626	-5.12**	-9.62**	-10.39**	-14.64**	18.35**	-22.22**
14.	EC-112112 x IC-128076	42.59**	27.86**	26.94**	13.83**	15.66**	3.72**
15.	EC-305626 x IC-	-4.85	11.17	-10.65	4.39	-27.63**	-15.45**

* Significant at 5 per cent level; ** Significant at 1 per cent level
(Source: Eswaran, 2007)

The extent of heterosis over mid-parent (relative heterosis) and better parent (heterobeltiosis) was significant in 60 and 53, out of 60 three way cross hybrids, respectively, for fruit yield per plant. The relative heterosis was highest with (Pusa Sawani × IC 128076) × EC 305626 followed by (EC 112112 × IC 128076) × Pusa Sawani and (Pusa Sawani × IC 128076) × Parbani Kranti. The three way cross combinations *viz.*, (Pusa Sawani × IC 128076) × EC 305626; (EC 305626 × IC 128076) × Pusa Sawani and (EC 112112 × IC 128076) × Pusa Sawani portrayed high better parent heterosis for fruit yield per plant. These three way cross combinations also exhibited high better parent heterosis for number of fruits per plant and fruit weight (Eswaran, 2007) (Table 68).

Table 68 : Heterosis in three way cross hybrids for fruit yield per plant

Three way cross hybrids	Fruit yield per plant	
	Relative heterosis (d_i)%	Heterobeltiosis (d_{ii}) %
AA x PK x PS	27.36**	10.89
AA x PK x EC- 112112	18.73**	7.73
AA x PK x EC-305626	16.74**	2.37
AA x PK x IC-128076	78.09**	50.89**
AA x PS x PK	31.14**	14.18
AA x PS x EC-112112	80.62**	52.44**
AA x PS x EC-305626	74.56**	42.01**
AA x PS x IC-128076	16.19*	-8.92*
AA x EC-112112xPK	40.31**	27.31**
AA x EC-112112 x PS	82.77**	54.25**
AA x EC-112112 x EC-305626	55.58**	32.29**
AA x EC-112112 x IC-128076	66.12**	36.32**
AA x EC-305626 x IC-128076	125.48**	78.16**
AA x IC-128076 x PK	33.77**	13.34*
AA x IC-128076 x PS	75.37**	37.47**
AA x IC-128076 x EC-112112	128.62**	87.61**
AA x IC-128076 x EC-305626	92.52**	52.12**
PK x PS x AA	39.47**	21.43**
PK x PS x EC-112112	51.22**	36.10**
PK x PS x EC-305626	62.62**	40.82**
PK x PS x IC-128076	150.17**	108.48**
PK x EC-112112 x AA	74.83**	58.63**
PK x EC-112112 x PS	90.53**	71.49**
PK x EC-112112 x EC-305626	105.22**	86.14**
PK x EC-112112 x EC-305626	81.48**	58.63**
PK x EC-305626 x AA	52.22**	33.49**
PK x EC-305626 x PS	80.96**	56.76**
PK x EC-305626 x EC-112112	23.98**	12.45
PK x EC-305626 x IC-128076	73.74**	46.00**
PK x IC-128076 x AA	86.63**	58.13**
PK x IC-128076 x PS	46.26**	21.89**
PK x IC-128026 x EC-112112	61.89**	41.51**
PK x IC-128076 x EC-305626	67.39**	40.67**

PS x EC-112112 x AA	70.51**	43.90**
PS x EC-112112 x PK	49.34**	34.41**
PS x EC-112112 x EC-305626	93.27**	77.46**
PS x EC-112112 x IC-128076	109.45**	84.76**
PS x EC-305626 x AA	109.47**	70.41**
PS x EC-305626 x PK	63.13**	41.32**
PS x EC-305626 x EC-112112	77.76**	63.22**
PS x EC-305626 x IC-128076	113.79**	103.26**
PS x IC-128076 x AA	107.41**	62.58**
PS x IC-128076 x PK	132.17**	93.58**
PS x IC-128076 x EC-112112	102.67**	78.77**
PS x IC-128076 x EC-305626	149.50**	137.20**
EC-112112 x EC-305626 x AA	90.38**	61.87**
EC-112112 x EC-305626 x PK	80.37**	63.60**
EC-112112 x EC-305626 x PS	100.42**	84.03**
EC-112112 x EC-305626 x IC-128076	114.54**	90.89**
EC-112112 x IC-128076 x AA	105.86**	68.93**
EC-112112 x IC-128076 x PK	109.87**	83.45**
EC-112112 x IC-128076 x PS	139.73**	111.47**
EC-112112 x IC-128076 x EC-305626	122.54**	98.01**
EC-305626 x IC-128076 x AA	86.76**	47.57**
EC-305626 x IC-128076 x PK	43.90**	20.93**
EC-305626 x IC-128076 x PS	130.20**	118.85**
EC-305626 x IC-128076 x EC-112112	96.56**	74.90**

* Significant at 5 per cent level; ** Significant at 1 per cent level.
(Source: Eswaran, 2007)

Chapter 7

Genetic Analysis of F_1 Generation

Eswaran (2007) made genetic analysis of six parents and their 30 hybrids of Okra with the method of analysis proposed by Hayman (1954b). He estimated the various genetic parameters *viz.*, $\hat{D}$, $\hat{F}$, $\hat{H}_1$, $\hat{H}_2$ and $\hat{E}$ as well as the various genetic ratio. The results are presented in Tables 69 and 70.

The estimates of $\hat{D}$ were significant for seven out of ten characters studied. This indicated that the component of variation due to additive genetic variance was important for days to seedling emergence, number of first fruiting node, height of first fruiting node, plant height, days to first picking, number of fruits per plant and fruit weight. These characters could be improved by resorting to simple selection. The estimates of $\hat{H}_1$ and $\hat{H}_2$ were positive and significant for almost all the characters studied. It indicated that there were unequal frequency of alleles i.e., u [1] v at all the loci, in this context, 'u' refers to the frequency of alleles which increase the mean expression of the character and are situated at loci which exhibited dominance. On the other hand, 'v' corresponds to the frequency of alleles at loci that decreases the expression of the character. Further proof for the unequal distribution of the alleles over loci was obtained by the ratio of $H_2/4H_1$ which was the estimate of $\overline{uv}$. In the present study, $\overline{uv}$ estimates were in the range of 0.14 to 0.23, which is less than the maximum value of 0.25. The maximum value of 0.25 would arise, when u=v = 0.5. If u=v, then

$\hat{H}_1 = \hat{H}_2$ i.e., the increaser and decreased alleles are in equal proportion in the parents. In the present study, $\hat{H}_1 \neq \hat{H}_2$

Table 69 : Estimates of genetic parameters for fruit yield and its component characters in Okra

S. No.	Characters	$\hat{D}$	$\hat{F}$	$\hat{H}_1$	$\hat{H}_2$	$\hat{h}^2$	$\hat{E}$
1.	Days to seedling emergence	0.34 ± 0.03**	0.32 ± 0.07**	0.21 ± 0.07*	0.13 ± 0.06*	0.05 ± 0.04	0.01 ± 0.01
2.	Days to first flowering	2.23 ± 1.12	3.59 ± 2.74	15.49 ± 2.85**	9.99 ± 2.55**	3.98 ± 1.71*	0.48 ± 0.42
3.	Number of branches per plant	0.69 ± 0.48	0.68 ± 1.17	4.92 ± 1.22**	3.90 ± 1.09**	0.68 ± 0.73	0.03 ± 0.18
4.	Number of first fruiting node	1.01 ± 0.31*	1.85 ± 0.78*	3.34 ± 0.80**	1.97 ± 0.72*	0.17 ± 0.49	0.02 ± 0.12
5.	Height of first fruiting node	24.06 ± 8.15*	12.09 ± 19.90	61.00 ± 20.68*	55.49 ± 18.47*	5.08 ± 12.43	1.16 ± 3.08
6.	Plant height	149.27 ± 9.13**	41.96 ± 23.30	146.37 ± 23.17**	130.90± 20.70**	-4.61 ± 13.93	0.37 ± 0.45
7.	Days to first picking	4.95 ± 2.05*	7.46 ± 5.01	30.64 ± 5.21**	20.18 ± 4.65**	8.02 ± 3.13*	0.22 ± 0.78
8.	Number of fruits per plant	22.90 ± 6.68**	1.05 ± 16.33	52.80 ± 16.97**	46.91 ± 15.16*	8.70 ± 10.20	0.79 ± 2.53
9.	Fruit weight	5.72 ± 1.73*	9.55 ± 4.22*	19.57 ± 4.39**	14.01 ± 3.92**	1.62 ± 2.64	0.44 ± 0.65
10.	Fruit yield per plant	4200.21 ± 5159.27	-3882.87 ± 12604.09	44436.84 ± 13097.27	39437.45 ± 11700.12**	10563.52 ± 7874.95**	73.89 ± 1950.02

* Significant at 5 per cent level; ** Significant at 1 per cent level. (Source: Eswaran, 2007)

The estimates of $\overline{uv}$ were very low i.e., 0.14 to 0.23. This indicated that the positive and negative alleles at the loci exhibiting dominance were not in equal proportion of the parents of interest. However, this estimate did not permit a determination as to which type of alleles occurred more frequently. The unequal distribution of positive and negative alleles indicated the operation of non-allelic interaction i.e., epistasis, through Vr, Wr and Wr, Wr′ graphs appeared to be non reliable. Therefore, the observed low regression could be the result of gene frequency other than 0.5 instead of epistasis. The theoretical consideration of Coughtrev and Mather (1970) and the computer stimulation of Feyt (1976) also indicated that Hayman (1954) test for detecting epistasis is reliable only if the genes are distributed independently in the parents involved in the diallel cross. These results made to think whether the observed heterosis is due to dominance or over dominance.

Table 70 : Ratios of genetic parameters for fruit yield and its component characters in Okra

S. No.	Characters	$(H_1/D)^{1/2}$	$H_2/4H_1$	$\frac{(4DH_1)^{1/2}+F}{(4DH_1)^{1/2}-F}$	h^2/H_2	Heritability in narrow sense (per cent)
1.	Days to seedling emergence	0.80	0.14	4.01	-0.04	13.00
2.	Days to first flowering	2.63	0.16	1.88	0.39	25.00
3.	Number of branches per plant	2.68	0.19	1.45	0.17	23.00
4.	Number of first fruiting node	1.82	0.15	3.02	0.08	10.00
5.	Height of first fruiting node	1.59	0.23	1.37	0.09	25.00
6.	Plant height	0.99	0.22	1.33	-0.03	45.00
7.	Days to first picking	2.49	0.16	1.87	0.39	24.00
8.	Number of fruits per plant	1.52	0.22	1.03	0.18	52.00
9.	Fruit weight	1.85	0.18	2.65	0.11	06.00
10.	Fruit yield per plant	3.25	0.22	0.75	0.27	52.00

(Source: Eswaran, 2007)

The estimates of $\hat{F}$ was positive and significant for days to seedling emergence, number of first fruiting node and fruit weight. It indicated the predominance of dominant alleles in the parent for these traits, in this generation. The estimates of $\hat{F}$ was positive but non-significant for days to first flowering, number of branches per plant, height of first fruiting node, plant height, days to first picking and number of fruits per plant. This may indicate that the parents of interest had more recessive alleles for these characters. For fruit yield per plant, the estimate of $\hat{F}$ was negative, implying that both dominance and recessive genes contributed equally to this trait, in the parents of present study. In fact the observed ratio of $\frac{(4DH_1)^{1/2} + F}{(4DH_1)^{1/2} - F}$ almost confirmed the above mentioned trend.

The estimates of $\hat{h}^2$ were positive and significant for days to first flowering, days to first picking and fruit yield per plant. It indicated the dominance effect expressed as the algebraic sum over all loci as heterozygous phases in all the crosses.

The degree of dominance averaged over all loci $(H_1/D)^{1/2}$ was less than unity for days to seedling emergence, indicating the presence of incomplete dominance for this trait, where as the regression line of Vr, Wr indicated complete dominance for this trait. For the remaining traits, except plant height, the ratio was more than unity. It indicated the presence of over dominance for these characters. Plan height was controlled by almost complete dominance. It may be pointed out that Vr, Wr graph for number of fruits per plant and fruit yield per plant indicated the presence of incomplete dominance, whereas the ratio of $(H_1/D)^{1/2}$ indicated the presence of over dominance for these traits. The difference in the gene action inferred from Vr, Wr graph and degree of dominance might be caused either due to presence of correlated gene distribution and / or due to unidirectional dominance with plus or minus effects and seemed to be not due to epistasis. It appeared that the additive and dominance components (Hayman, 1954b) may likely to be confounded due to the presence correlated gene distribution. Therefore, both the assumptions i.e., "no epistasis" and "no correlated gene distribution" underlying diallel analysis seemed to be not valid (Jagtap and Kolhe, 1987).

The estimates of number of effective factors (h^2/H_2) were less than unity for all the characters studied. This under estimation may be due to the fact that the dominance effects of the genes affecting these traits are not equal in size and direction or if the distribution of the genes is correlated (Mather, 1949 and Jinks, 1954).

The narrow sense heritability estimates were high for plant height, number of fruits per plant and fruit yield per plant. These characters may be controlled by additive genetic variance. This indicated that the individual genotype can be evaluated readily from their phenotypic expression. Simple selection will be more effective in these sets of materials exhibiting greater additive genetic variability and desirable mean performance. Thus, it merits selection in the next generation. On the other hand, the remaining traits may be largely be controlled by non additive genetic variance or the number of genes controlling these traits may be more or these traits may be largely influenced by modifiers. The result indicated that the environmental effects were low and hence, the observed low heritability might be due to other factors other than environment. These traits are hard to improve by selection. It necessitates progeny testing.

Chapter 8

Graphic Analysis in F_1 Generation

Eswaran (2007) made graphic analysis with the model proposed by Jinks and Hayman (1953). He evolved 30 hybrids from six parents in a diallel fashion. The results are presented in Tables 71-74.

Table 71 : t^2 values for V_rW_r regression for fruit yield and its component characters in Okra

S. No.	Characters	t^2 value
1.	Days to seedling emergence	4.83
2.	Days to first flowering	0.46
3.	Number of branches per plant	0.96
4.	Number of first fruiting node	0.22
5.	Height of first fruiting node	0.01
6.	Plant height	0.01
7.	Days to first picking	0.74
8.	Number of fruits per plant	1.54
9.	Fruit weight	0.14
10.	Fruit yield per plant	6.28

(Source: Eswaran, 2007)

Table 72 : Estimates of Y intercept (a) and slope (b) of best fitting regression line for fruit yield and its component characters in Okra

S. No.	Characters	V_rW_r		W_rW_r	
		a	b	a	b
1.	Days to seedling emergence	-0.003	1.44±0.37	0.003	0.26±0.02
2.	Days to first flowering	-0.94	0.33±0.11	1.08	-0.01±0.02
3.	Number of branches per plant	-0.18	0.28±0.09	0.01	-0.04±0.01
4.	Number of first fruiting node	-0.35	0.63±0.17	0.01	-0.29±0.09
5.	Height of first fruiting node	6.26	0.15±0.16	3.11	0.15±0.01
6.	Plant height	0.96	0.94±0.15	1.25	0.46±0.08
7.	Days to first picking	-1.70	0.33±0.10	1.98	0.03±0.02
8.	Number of fruits per plant	6.98	0.23±0.08	1.59	0.48±0.12
9.	Fruit weight	-2.01	0.61±0.17	0.46	0.01±0.05
10.	Fruit yield per plant	1224.10	0.14±0.06	2523.46	0.24±0.07

(Source: Eswaran, 2007)

Table 73 : Significant deviation of 'b' for W_r/V_r and W_r¢/W_r for fruit yield and its component characters in Okra

S. No.	Characters	V_rW_r		W_rW_r	
		$\frac{(b-0)}{SE(b)}$	$\frac{(1-b)}{SE(b)}$	$\frac{(b-0)}{SE(b)}$	$\frac{(1-b)}{SE(b)}$
1.	Days to seedling emergence	3.89**	-1.19	9.55**	8.82**
2.	Days to first flowering	2.98*	6.05**	-0.40	20.41**
3.	Number of branches per plant	2.90*	7.33**	-3.16	46.80**
4.	Number of first fruiting node	3.58**	2.13*	-3.13	8.55**
5.	Height of first fruiting node	0.96	2.55**	8.68**	20.26**
6.	Plant height	6.26**	0.40	5.75**	0.50
7.	Days to first picking	3.24*	6.63*	1.20	18.87**
8.	Number of fruits per plant	2.55*	8.69**	4.10**	0.17
9.	Fruit weight	3.67*	2.34*	0.25	12.25**
10.	Fruit yield per plant	2.19*	13.38**	3.17**	3.43*

* Significant at 5per cent level ** Significant at 1per cent level
(Source: Eswaran, 2007)

Table 74 : Correlation coefficient 'r' for Y_r, $(W_r + V_r)$ for fruit yield and its component characters in Okra

S. No.	Characters	'r'
1.	Days to seedling emergence	-0.24
2.	Days to first flowering	0.79
3.	Number of branches per plant	0.07
4.	Number of first fruiting node	-0.11
5.	Height of first fruiting node	0.08
6.	Plant height	-0.06
7.	Days to first picking	0.73
8.	Number of fruits per plant	0.46
9.	Fruit weight	-0.22
10.	Fruit yield per plant	0.71

(Source: Eswaran, 2007)

Graphical analysis

t^2 values for Vr, Wr regression

The estimates of uniformity test were non-significant for all the ten characters studied. It indicated that the assumptions made by Hayman (1954b) were valid for all the ten traits of interest.

Deviation of 'b' for Vr, Wr and Wr Wr'

The deviation of regression coefficients from zero were significant for nine out of ten characters studied. It was non-significant for height of first fruiting node. It indicated that the characters other than height of first fruiting node had a strong relation ship between Vr (Variance of each array) and Wr (Co-variance between parents and their off springs) of the parental materials and thus Vr, Wr graph was effective for the genetical studies of the parental materials with respect to these attributes. For height of first fruiting node, the deviation of regression coefficient from zero was non-significant. It indicated that the relationship between Vr, Wr is non-significant. Hence, the Vr, Wr graph should not be plotted. It may be due to environmental factors, genotype x environmental interactions or sampling error (Singh *et al.*, 2001).

The regression co-efficient of Vr, Wr graphs were significantly different from unity for days to first flowering, number of branches per plant, number of first fruiting node, height of first fruiting node, days to first picking, number of fruits per plant and fruit yield per plant. It may indicate the involvement of epistasis for these characters. For the remaining characters the regression coefficients of Vr, Wr graph were non-significantly different from unity. It may indicate the involvement of additivity.

Estimates of Y intercept ('a') and slope ('b') of the best fitting regression line

The estimates of Y intercept ('a') close to zero for days to seedling emergence, days to first flowering, number of branches per plant, number of first fruiting node, days to first picking and fruit weight. This indicated the presence of complete dominance for these traits. For the remaining traits, it was positive. It may indicate the presence of partial dominance for these characters. The estimates of 'Y' intercept ('A') was positive for all the traits studied. This may indicate the involvement of partial dominance.

Correlation co-efficient 'r' for y r, (Wr + Vr) for ten characters

The correlation coefficients ('r') between standardized parental measurements (Yr) and parental order of dominance (Wr + Vr) were non-significant. It suggested that dominant as well as recessive genes controlled both high or low mean performance (Mather and Jinks, 1971).

Fruit yield per plant

The regression line of Vr, Wr intersected the Wr axis well above the origin. It indicated the presence of incomplete dominance. The relative value of Vr, Wr indicated that the parents P_3 and P_6 had most dominant genes, where as the parent P_2 had most recessive genes. The remaining parents $P_{1,}$ P_4 and P_5 had both dominant and recessive genes.

The proximity of the points P_1, P_4 and P_5 from each other suggested that these parents have similar genotypes and that differences among them might be due to genes with relatively small total influence on this character. Similarly, the points P_3 and P_6 represents another group of parents having similar genotypes. The parent P_2 alone formed a separate group. The result indicated the presence of sufficient amount of genetic diversity among the parents.

The standard deviation graph Yr, (Wr +Vr) indicated that the parents P_1 and P_2 had recessive genes with positive effects, where as the parent P_5 had recessive genes with negative effect. The parents P_3 and P_6 had dominant genes with negative effect, where as the parent P_4 had dominant genes with positive effect. The over all information derived from Vr, Wr; Wr Wr and standard deviation graph Yr, (Wr +Vr) is furnished below.

The regression line of the Vr, Wr graph intersected the Wr axis was exactly at the origin for days to seeding emergence and plant height. It indicated that these traits were controlled by complete dominance. The regression line of the Vr, Wr graph intersected the Wr axis well below the origin for days to first flowering, number of branches per plant, number of first fruiting node, days to first picking and fruit weight. It indicated the presence of over dominance in the inheritance of these traits. The regression line of Vr, Wr intersected the Wr axis well above the origin for number of fruits per plant and fruit yield per plant. It indicated the presence of incomplete dominance in the inheritance of these traits. Incomplete dominance can be had as a case of semi-additivity.

The relative value of Vr, Wr showed that the high yielding parent P_1 (AA) had both dominant and recessive genes equally for days to seedling emergence, number of branches per plant, and number of first fruiting node. It had most recessive genes for remaining traits. The high yielding parent did not have dominant genes.

The mean fruit yield per plant was maximum with PK x AA, AA x PK and PK x EC-305626. The first cross had female parent with both dominant and recessive genes and the male parent had recessive genes. The reverse was true for the next high yielding cross AA x PK. The next high yielding cross PK x EC-305626 had both the parents with dominant as well as recessive genes. In general, the distribution of genes for fruit yield was found to be asymmetrical. This asymmetry follows neither high nor low yield (Hayman, 1958).

In addition to Vr, Wr graph, the standard deviation graph indicated the direction of the effects of dominance and recessive genes in the parents. The position of array points of many parents for many traits in Vr, Wr graph, however, were not in accordance with the dominance and recessiveness of the parents in Yr, (Wr + Vr) graph. These anomalies indicated that these parents might be involved in non-allelic interaction to a greater extent (Jagtap and Kolhe, 1987).

The hybrids of the cross combinations *viz*., AA x EC-112112, EC-112112 x AA, AA x EC-305626, EC-305626 x AA, AA x IC-128076, IC-128076 x AA, PK x PS, PS x PK, EC-305626 x PS, PS x IC-128076, IC-128076 x PS, EC-112112 x EC-305626, EC-305626 x EC-112112 x EC-305626 x IC-128076 and IC-128076 x EC-305626 were the cross combinations, which recorded negative standard heterosis for fruit yield per plant. Similarly, the cross combinations *viz*., PK x IC-128076, IC-1128076 x PK, PS x EC-112112, EC-112112 x PS, PS x EC-305626 and EC-305626 x PS were the cross combinations which had not exhibited significant positive standard heterosis for fruit yield per plant. The parents involving these cross combinations had recessive genes with positive effect or recessive genes with negative effect or dominant genes with positive effect or dominant genes with negative effect, as per standard deviation graph. As per Vr, Wr graph, they had most recessive genes or most dominant genes or equal frequency of dominant and recessive genes. This might indicate that dominant or recessive genes are not the cause for heterosis. But, the combining ability might be the cause for heterosis. In fact, the parents of the aforementioned cross combinations mostly had either non significant *gca* effects. Similarly, the cross combinations involving the above said parents showed non-significant or negative significant *sca* effects. Interestingly, the cross combinations which recorded high mean and standard heterosis for fruit yield per plant had parents (at least one) mostly, with positive significant *gca* effects and their cross combination had positive significant *sca* effects, mostly. Hence, it may be pointed out that the combining ability of the parents could be the cause for heterosis and may not be the gene distribution.

The correlation co-efficients ('r') between standardized parental measurements (Yr) and parental order of dominance (Wr + Vr) were non-significant for all the characters studied. It suggested that dominant as well as recessive genes controlled both high and low mean performance. It indicated the presence of amphi-directional dominance (Mather and Jinks, 1971).

In Vr, Wr graph for fruit yield per plant, the proximity of the points P_1 (AA), P_4 (EC-112112) and P_5 (EC -305626) indicated that these parents may have near to identical genotypes. It formed a separate group. Similarly, the parent P_3 (PS) and P_6 (IC-128076) were close to each other and formed a separate group. The parent P_2 (PK) alone formed a separate group. The results indicated that there were sufficient amount of genetic diversity among the parents of interest.

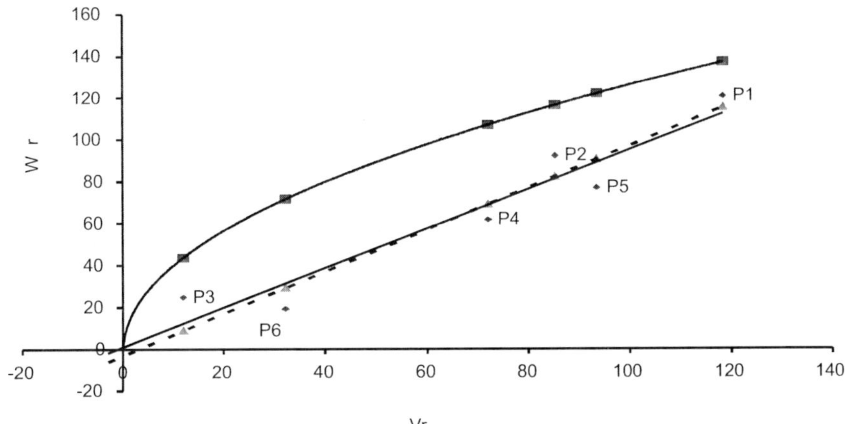

Fig. 1a : Vr, Wr graph for plant height

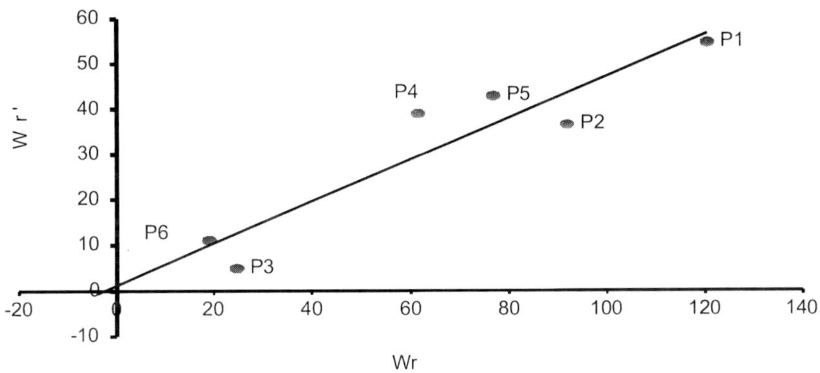

Fig. 1b : Wr, Wr graph for plant height

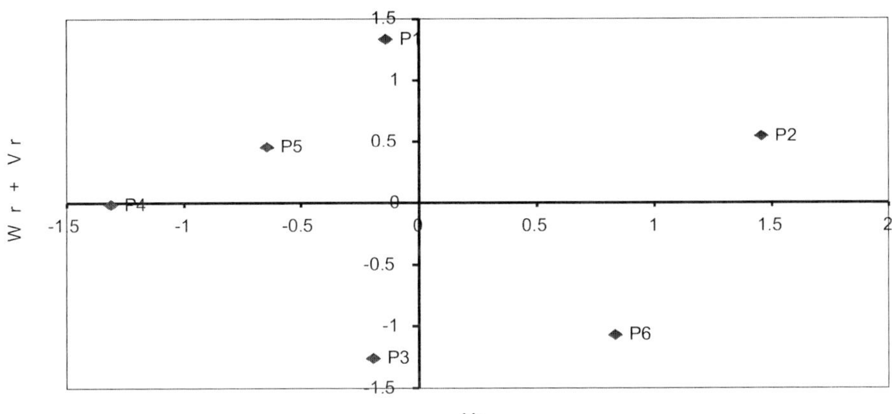

Fig. 1c : Yr, (Wr + Vr) graph for plant height
(Source : Eswaran, 2007)

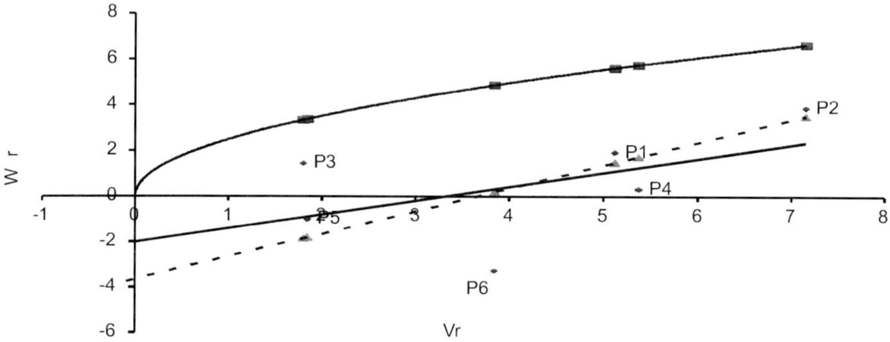

Fig. 2a : Vr, Wr graph for fruit weight

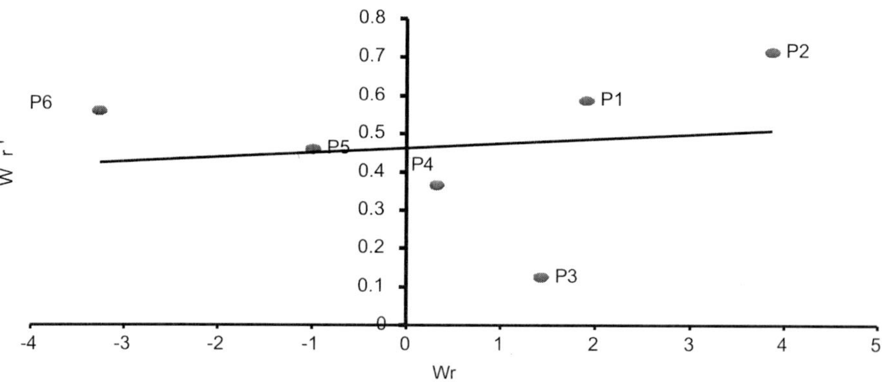

Fig. 2b : Wr, Wr graph for fruit weight

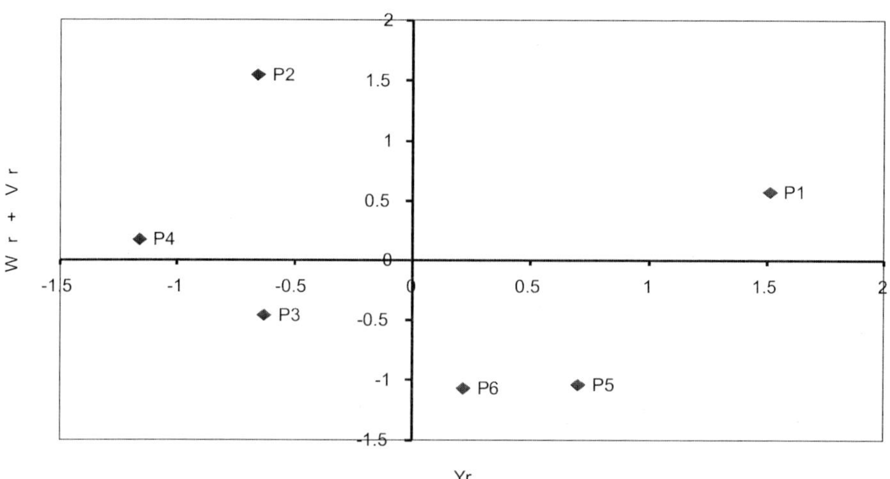

Fig. 2c : Yr, (Wr + Vr) graph for fruit weight
(Source : Eswaran, 2007)

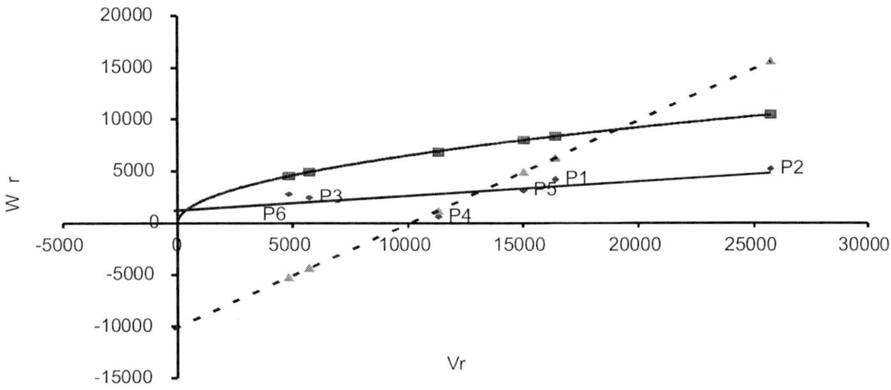

Fig. 3a : Vr, Wr graph for fruit yield per plant

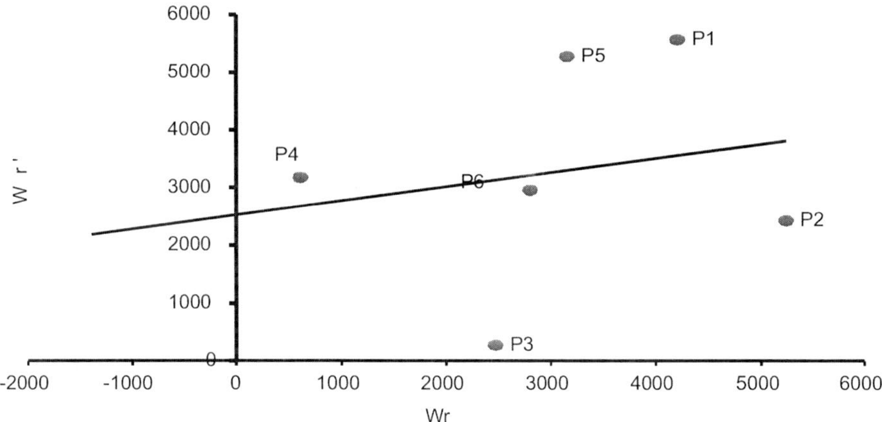

Fig. 3b : Wr, Wr graph for fruit yield per plant

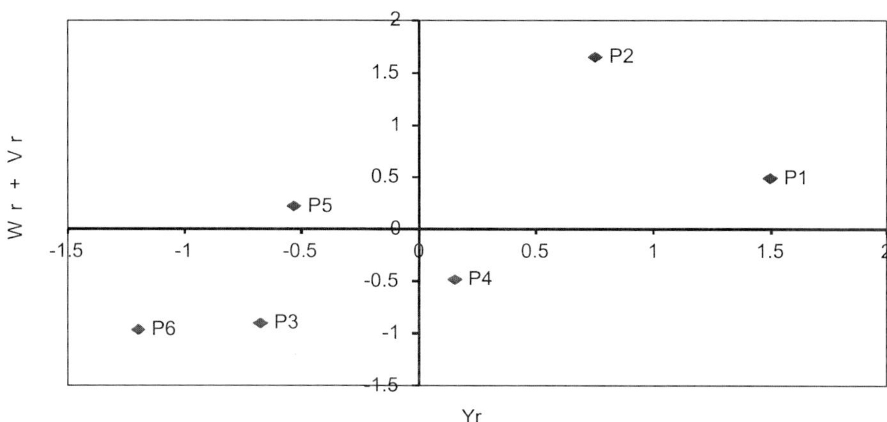

Fig. 3c : Yr, (Wr + Vr) graph for fruit yield per plant
(Source : Eswaran, 2007)

The cross combinations which recorded maximum fruit yield per plant *viz.*, PK x AA, AA x PK and PK x EC-305626 possessed parents completely from different groups. This indicated that the observed heterosis might be partly due to genetic diversity among the parents used in the present investigation. Thus, the parents included in the present study offered good scope for their inclusion in heterosis breeding programme. Their combinations would offer very good scope for further development of different traits.

Moreover, the parents which showed high *gca* effects for fruit yield per plant *viz.*, AA (P_1), PK (P_2) and EC-112112 (P_4) were mostly of diverse origin. The cross combinations which recorded highest over all score for *sca* effects namely, AA x PK and PS x EC-112112 had parents from different genetic groups. Similarly, the reciprocal cross combinations which recorded highest over all score for *rca* effects namely, PK x AA, PS x PK, PS x AA, EC-305626 x PK, IC-128076 x EC -112112 mostly had parents from different genetic groups. Thus, a difficulty was experienced to separate the effects of general genetic diversity and the gene effects on *sca* effects. It becomes open to question. Probably, the genetically diverse parents may have more combining ability than that of closely related ones. A difficulty was also experienced to separate the effects of heterozygosity on the heterotic performance, as they are confounded and becomes open to next question.

The present study indicated that only Vr, Wr and Wr, Wr′ graphs are not providing sufficient information regarding the genetic make up of the parents. Because dominant as well as recessive genes also expressed both high and low performance.

Genetic parameters

The additional genetic statistics needed for the genetic interpretation using Hayman (1954b).diallel analysis are given below.

Days to seedling emergence

The estimate of $\hat{D}$ was significant for this character. It indicated that the component of variation due to additive effects of the genes influenced this character. The estimate of was positive and significant. It indicated the presence of more dominant alleles than the recessive alleles. The estimates of $\hat{H}_1$ and $\hat{H}_2$ were significant. It indicates the presence of dominant genetic variances. The estimate of h_2 was significant, it implied the dominance effect expressed as the algebraic sum over all loci in heterozygous phase in all crosses. The estimate of was non-significant.

The potence ratio was less than the unity indicating the presence of incomplete dominance. Incomplete dominance can be had as a case of additivity. The ratio of $H_2/4H_1$ was less than 0.25, it indicated the unequal distribution of the positive and negative alleles for this character. The ratio of $\frac{(4DH_1)^{1/2}+F}{(4DH_1)^{1/2}-F}$ was more than one. It indicated that the parents carried more dominant alleles for this character. There was under estimation of the number of effective factors. It may be due to the fact that dominant effects of the genes affecting this trait are not in equal size and direction or if the distribution of genes is correlated (Jinks, 1954; Mather, 1949). Narrow sense heritability estimates was lower, implying the involvement of non-additivity in the expression of this trait.

Chapter 9

Genetic Analysis of Early Segregating Generations

The narrow sense heritability estimates were high for days to first flowering, plant height at first flowering, number of leaves at first flowering, plant height at maturity, productive tillers per plant, boot leaf area, 100 grain weight, grain length and grain L/B ratio. These characters may be controlled by additive genetic variance. This indicated that the individual genotype can be evaluated readily from their phenotypic expression. Simple selection would more effective in the sets of materials exhibiting greater additive genetic variability and desirable mean performance. Thus, it merit selection in the next generation. On the other hand, the remaining traits may largely be controlled by non-additive genetic variance or the number of genes controlling these traits may be more, or these traits may largely be influenced by a large number of modifiers or might be largely influenced by environment. As the effects of the environmental variance were generally low, the observed low heritability for the above traits might be due other factors other than environment. These traits are hard to improve by selection. It necessitates progeny testing.

9.1. Generation mean analysis

The development of plant breeding strategy hinges mainly on the support provided by the genetic information on the inheritance and behaviour of major quantitative characters associated with the yield and other economic traits concern associated with the yield and other economic traits concern to the breeder (Arunachalam, 1976). An insight into the magnitude of variability present in the breeding material is of immense importance as it provides the basis for effective selection. The breeders are mostly concerned with the segregating populations like F_2, F_3, etc. derived from crosses involving widely different genotypes for their selection programmes. Therefore, identification of appropriate parents and combining them into hybrids in a planned way from the most important prerequsite for a selection programme.

Fisher in 1918, grouped the genotypic variance into additive, dominance and epistatic interaction between the two homozygotes of a gene, AA and aa. Dominance component is due to the deviation of the heterozygote (Aa) from the averages of the two homozygotes (AA and aa). The epistatic component results from the interaction between two or more genes. Later, Hayman and Mather (1955) partitioned the epistatic component into three types of interactions viz., additive × additive, additive × dominance and dominance × dominance. For selecting appropriate breeding procedure for the improvement of economic characters it is essential to draw precise informations on the various types of gene effects so that the crop improvement programme will be very efficient and required result could be achieved.

Several procedures have been outlined for the prediction of gene action using either the mean values of different generations on the variances calculated for various types of populations. The generation mean analysis is considered to be one of the best methods for estimating the different component of variance. A set of segregating and non-segregating generation belonging to a common ancestry is subjected to analysis for estimating additive, dominance and epistatic components of variances simultaneously. Each cross is delt as a unit and estimates were obtained for each cross. In this way, generation mean analysis serves as a very precise method recommended for dealing with individual crosses.

This part of the present investigation, generation mean analysis involving a five parameters model was applied to estimate genetic components including epistasis for sixteen traits of 6 cross combination in rice.

9.2. Statistical analysis

Residual heterosis in F_2 and F_3 were estimated from generation mean of the crosses and expressed as per cent over mid parent (di-relative heterosis). The mean of F_2 and F_3 generations over replication were utilised for the estimation of residual heterosis.

Residual heterosis in F_2 generation

$$di = \frac{\overline{F_2} - \overline{MP}}{\overline{MP}} \times 100$$

$\overline{F_2}$ = mean of the F_2 generation

$\overline{MP}$ = mean of the mid parental value

Residual heterosis in F_3 generation

$$di = \frac{\overline{F_3} - \overline{MP}}{\overline{MP}} \times 100$$

$\overline{F_3}$ = mean of the F_3 generation

$\overline{MP}$ = mean of the mid parental value

9.2.1. Test of significance for residual heterosis

The significant of residual heterosis in F_2 and F_3 were tested using the formula suggested by Wynne *et al.* (1970).

$$\text{'t' over (di)} = \frac{\overline{F_2} - \overline{MP}}{(3\sigma^2 e / 2r)^{1/2}}$$

$$\text{'t' over (di)} = \frac{\overline{F_3} - \overline{MP}}{(3\sigma^2 e / 2r)^{1/2}}$$

where

$\sigma^2 e$ = is error variance derived from the analysis of variance

r = number of replication

9.2.2. Inbreeding depression

9.2.2.1. Estimation of inbreeding depression

The inbreeding depression was worked out in F_2 and F_3 as per the formula given below.

Inbreeding depression in F_2

$$di = \frac{\overline{F}_1 - \overline{F}_2}{\overline{F}_1} \times 100$$

$\overline{F}_1$ = mean of the F_1 generation

$\overline{F}_2$ = mean of the F_2 generation

In breeding depression in F_3

$$di = \frac{\overline{F}_1 - \overline{F}_3}{\overline{F}_1} \times 100$$

$\overline{F}_1$ = mean of the F_1 generation

$\overline{F}_2$ = mean of the F_3 generation

9.2.2.2. Test of significance

The significance were tested using the formulae suggested by Wynne *et al.* (1970).

$$\text{'t' over } (F_2) = \frac{\overline{F}_1 - \overline{F}_2}{(3\sigma^2 e / 2r)^{1/2}}$$

$$\text{'t' over } (F_3) = \frac{\overline{F}_1 - \overline{F}_3}{(3\sigma^2 e / 2r)^{1/2}}$$

where,

$\sigma^2 e$ = is error variance derived from the analysis of variance

r = number of replication

9.2.3. Generation mean analysis

The mean values were computed for all the five generations viz., P_1, P_2, F_1, F_2 and F_3 over all replications for each cross. The variance and the variance of mean were also computed for each of the five generations for three cross combinations.

9.2.3.1. Scaling test

The means of the different generations were utilized for obtaining the various genetic effects. The adequacy of the data for a simple additive - dominance model was tested utilizing the scales C and D as suggested by Mather (1949). The C and D scales were calculated by using the following formulae.

Scale C = $4\bar{F}_2 - 2\bar{F}_1 - \bar{P}_1 - \bar{P}_2 = 0$

Scale D = $4\bar{F}_3 - 2\bar{F}_2 - \bar{P}_1 - \bar{P}_2 = 0$

where, $\bar{P}_1$, $\bar{P}_2$, $\bar{F}_1$, $\bar{F}_2$ and $\bar{F}_3$ are the means of different generations over all the replications

$V_C = 16V_{\bar{F}_2} + 4V_{\bar{F}_1} + V_{\bar{P}_1} + V_{\bar{P}_2}$

$V_D = 16V_{\bar{F}_3} + 4V_{\bar{F}_2} + V_{\bar{P}_1} + V_{\bar{P}_2}$

where, V_C and V_D are the variances of the scales C and D and $V_{\bar{P}_1}$, $V_{\bar{P}_2}$, $V_{\bar{F}_1}$, $V_{\bar{F}_2}$ and $V_{\bar{F}_3}$ are the variances of mean of P_1, P_2, F_1, F_2 and F_3 generations respectively.

The standard errors of C and D were obtained as $\sqrt{V_C}$ and $\sqrt{V_D}$ respectively and were utilized for testing the significance of the deviations of the scales from zero. The significance of scales C and D was determined by comparing the calculated and table 't' values.

Calculated 't' for C = $\dfrac{C}{\sqrt{V_C}}$

Calculated 't' for D = $\dfrac{D}{\sqrt{V_D}}$

The additive-dominance model was considered inadequate when any of the scales was found to deviate significantly from zero.

9.2.3.2. Estimation of genetic effects

In cases where the scales C or D significantly differed from zero, assuming a digenic interaction model, five parameters viz., (m), (d), (h), (i) and (l) were estimated (Hayman, 1958).

(m) = $\bar{F}_2$

(d) = $½\,\bar{P}_1 - ½\bar{P}_2$

(h) = $1/6\left(4\bar{F}_1 + 12\bar{F}_2 - 16\bar{F}_3\right)$

(i) = $\bar{P}_1 - \bar{F}_2 + ½\,(\bar{P}_1 - \bar{P}_2 + h) - ¼\,1$

(l) = $1/3\,(16\bar{F}_3\text{-}24\bar{F}_2\text{+}8\bar{F}_1)$

where,

(m) = mean of F_2

(d) = additive effect

(h) = dominance effect

(i) = additive × additive interaction effect

(l) = dominance × dominance interaction effect

The variance for (m), (d), (h), (i) and (l) were estimated as follows:

$$V_{(m)} = V_{\bar{F}_2}$$

$$V_{(d)} = ¼\left(V_{\bar{P}_1} + V_{\bar{P}_2}\right)$$

$$V_{(h)} = 1/36\left(16V_{\bar{F}_1} + 144V_{\bar{F}_2} + 256V_{\bar{F}_3}\right)$$

$$V_{(i)} = V_{\bar{P}_1} + V_{\bar{P}_2} + ¼\left(V_{\bar{P}_1} + V_{\bar{P}_2} + V_h\right) + 1/16\ V_1$$

$$V_{(l)} = 1/9\left(256V_{\bar{F}_3} + 576V_{\bar{F}_2} + 64V_{\bar{F}_1}\right)$$

The standard error was calculated from the respective variances as follows:

SE for (m) = $\sqrt{V_m}$

SE for (d) = $\sqrt{V_d}$

SE for (h) = $\sqrt{V_h}$

SE for (i) = $\sqrt{V_i}$

SE for (l) = $\sqrt{V_l}$

The significance of the parameters was tested by calculating the 't' values as done for the scaling test.

The data were transformed into arc sine transformation whenever the data registered below 30 per cent and above 70 per cent.

9.3. Example

Narasimman (2005) evolved six crosses ($P_1 \times P_2$); ($P_2 \times P_1$); ($P_1 \times P_3$); ($P_3 \times P_1$); ($P_2 \times P_3$); ($P_3 \times P_2$) in rice with three parents *viz.*, ADT 37 (P_1), ADT 38 (P_2) and ADT 44 (P_3).

9.3.1. Mean performance

Several procedures have been outlined for the prediction of gene action using either the mean values of different generations on the variances calculated for various types of populations. The generation mean analysis is considered to be one of the best methods for estimating the different component of variance. A set of segregating and non-segregating generation belonging to a common ancestry in subjected to analysis for estimating additive, dominance and epistatic components of variances simultaneously. Each cross is delt as a unit and estimates were obtained for each cross. In this way, generation mean analysis serves as a very precise method recommended for dealing with individual crosses.

This part of the present investigation, generation mean analysis involving a five parameters model was applied to estimate genetic components including epistasis for sixteen traits of 6 cross combination in rice. The prime objectives of hybridisation between any two parents is to combine the desirable characters dispersed among them to compensate the deficiencies found in one parent by the other. The progenies of these crosses will give all possible combinations to promote yield. Therefore, the parents selected should inherit the characters to their progenies. The choice of parent is based on the general principle that the parents under selection should have a high *per sc* performance for the desirable traits. For a systematic breeding programme, it is necessary to identify the parents which can be exploited for genetic improvement through hybrid progenies. Breeders are in absolute need of high mean value as a main criteria for effective selection.

The parent P_1 flowered earlier than the other parents. The F_1 hybrids of the cross $P_1 \times P_2$ flowered earlier than the other hybrids. But their F_2 and F_3 progenies flowered lately. The flowering was earlier with F_2 and F_3 progenies of the cross P_2 P_3. Similarly the F_3 progenies of $P_2 \times P_1$ flowered earlier. The result indicated that early × medium, medium × early and medium × late cross combinations flowered earlier. Selection among this progenies is likely give some useful early segregants. The F_2 progenies of the cross $P_2 \times P_3$ and F_3 progenies of the crosses $P_2 \times P_1$ and $P_2 \times P_3$ demonstrated higher mean grain yield per plant. The F_2 and F_3 progenies of the cross $P_2 \times P_3$ had higher grain length. The F_2 and F_3 progenies of this cross combination also showed higher L/B ratio. The partitioning efficiency of the F_2 and F_3 progenies of this cross ($P_2 \times P_3$) was also considerable. So the cross combination involving the parents P_2 and P_3 (medium × late) yielded early maturing, high yielding genotypes with superior grain quality.

The F_1, F_2 and F_3 progenies of the cross combination viz., $P_3 \times P_1$ were high yielding. Its reciprocal cross was also rewarding. The F_1, F_2 and F_3 progenies of the cross $P_3 \times P_1$ also recorded medium duration for first flowering. The F_3 progenies of this cross combination recorded higher grain length and desirable L/B ratio. The partition efficiency of these progenies was also considerable.

The result clearly indicated the superiority of the cross combinations viz., $P_2 \times P_3$ (medium × late) and $P_3 \times P_1$ (late × early) in the earliness breeding programme coupled with high yield, desirable grain quality and partitioning efficiency. Potential segregants could also be derived from these crosses. Infact, their segregating progenies demonstrated increased performance than their parents, which amply indicated the presence of transgressive segregation in the desirable direction. According to Chawla and Gupta (1983), the parents with high *per se* performance could produce transgressive segregants in F_2 as well as later generations.

Table 75 : Performance of parents F_1's, F_2's and F_3's – grain yield per plant

Parameters	Generations							
	Parent	Parent	F_1		F_2		F_3	
Cross: $P_1 \times P_2$	P_1	P_2	$P_1 \times P_2$	$P_2 \times P_1$	$P_1 \times P_2$	$P_2 \times P_1$	$P_1 \times P_2$	$P_2 \times P_1$
Range	6.10 (25.80 – 31.90)	6.22 (31.90 – 38.12)	9.70 (27.90 – 37.60)	12.09 (29.21 – 41.30)	36.83 (8.89 – 45.72)	27.10 (9.22 – 36.32)	17.00 (6.00 – 23.00)	27.88 (12.02 – 39.90)
Mean	28.74	34.87	32.70	34.76	17.44	17.61	15.14	19.26
S.D	-	-	-	-	5.72	5.21	3.20	5.55
CV (%)				-	32.80	29.59	27.73	28.81
Rel. H. (%)	-	-	2.81**	9.30**	-	-	-	-
Res. H. (%)	-	-	-	-	-45.17**	-44.63**	-49.17**	-39.44**
IB (%) (%)	-	-	-	-	46.67**	49.34**	50.56**	44.59**
Cross: $P_1 \times P_3$	P_1	P_3	$P_1 \times P_3$	$P_3 \times P_1$	$P_1 \times P_3$	$P_3 \times P_1$	$P_1 \times P_3$	$P_3 \times P_1$
Range	6.10 (25.80 – 31.90)	4.08 (36.51 – 40.59)	11.25 (31.70 – 42.95)	11.60 (34.80 – 46.40)	38.92 (8.68 – 47.60)	32.09 (10.61 – 42.70)	23.91 (11.89 – 35.80)	36.47 (12.24 – 48.71)
Mean	28.74	38.46	38.88	40.08	19.05	21.34	20.83	24.65
S.D	-	-	-	-	6.77	5.93	4.04	6.99
CV (%)	-	-	-	-	27.80	35.52	28.36	19.41
Rel. H. (%)	-	-	14.59**	19.28**	-	-	-	-
Res. H. (%)	-	-	-	-	-43.29**	-36.47**	-38.01**	-26.62**
IB (%) (%)	-	-	-	-	50.51**	46.74**	45.90	38.48**

Cross: $P_2 \times P_3$	P_2	P_3	$P_2 \times P_3$	$P_3 \times P_2$	$P_2 \times P_3$	$P_3 \times P_2$	$P_2 \times P_3$	$P_3 \times P_2$
Range	6.22 (31.90 - 38.12)	4.08 (36.51 - 40.59)	8.30 (35.10 - 43.40)	15.06 (31.10 - 46.16)	31.42 (10.06 - 41.48)	31.74 (5.88 - 37.62)	29.67 (13.30 - 42.97)	25.46 (10.34 - 35.80)
Mean	34.87	38.46	39.56	37.55	20.56	16.22	22.14	20.23
S.D	-	-	-	-	5.22	5.53	5.38	4.15
CV (%)	-	-	-	-	25.38	34.12	24.32	20.50
Rel. H. (%)	-	-	-7.91**	2.43**	-	-	-	-
Res. H. (%)	-	-	-	-	-43.93**	-55.75**	-39.61**	-44.81**
IB (%) (%)	-	-	-	-	48.04**	56.80**	44.03**	46.12**

* – Significant at 5 per cent level ** – Significant at 1 per cent level
(Source: Narasimman, 2005)
( ) – Figures in parenthesis indicate minimum and maximum value.

9.3.2. Genetic variability, heritability and genetic advance

The progress in breeding for a economic traits depends on the magnitude and nature of the genetic variability. The biological variation occurring in this crop of great of scope for genetic improvement through selection. The current interest in rice improvement is towards the development of early maturing varieties with high yield coupled with desirable grain quality. The potentiality of the cross is measured not only by mean performance but also the extent of variability created for economic characters.

Heritability value alone provides no indication on the amount of genetic progress. According to Hanson (1961), heritability and genetic advance are complementary aspects. Ramanujam and Tirumalachar (1967) discussed the limitation of estimating heritability in broad sense and suggested that heritability in broad sense is reliable, when it is accompanied by high genetic advance. In a planned crop breeding programme, efficient use of genetic variation present in the breeding material, always desirable to study the inheritance of economic traits of predicting the heritability and genetic advance in selection programme (Burton, 1952; Johnson *et al.,* 1955).

For grain yield per plant, the F_2 population of $P_1 \times P_3$ and the F_3 population of P_2´P_1 evinced high PCV and GCV coupled with high heritability and genetic advance. This indicated the presence of additive gene action. Hence, there is scope for selection. The influence of environment was less as inferred from the low difference between PCV and GCV.

The F_2 population of the cross $P_1 \times P_3$ which showed high PCV and GCV coupled with high heritability and genetic advance grain yield per plant also showed high genetic parameters for number of leaves at first flowering, filled grains per panicle, 100 grain weight and biomass per plant. The F_2 population of the cross $P_3 \times P_1$ which recorded high genetic parameters for days to first flowering also recorded high genetic parameters for plant height at maturity, spikelet sterility, grain length and grain L/B ratio. The result indicated that there is scope for evolving early maturing varieties with good grain quality and yield.

Table 76 : Variability heritability and genetic advance in segregating generation in residual rice-grain yield per plant

Parameters	Generations	$P_1 \times P_2$	$P_2 \times P_1$	$P_1 \times P_3$	$P_3 \times P_1$	$P_2 \times P_3$	$P_3 \times P_2$
PCV (%)	F_2	31.04	28.63	34.29	26.95	23.99	32.69
	F_3	26.31	28.30	18.56	27.55	23.43	19.05
GCV (%)	F_2	30.09	28.11	33.63	26.49	23.24	31.92
	F_3	25.54	28.02	18.11	27.11	22.96	18.26
Heritability (%)	F_2	93.99	96.42	96.19	96.62	93.86	95.37
	F_3	94.29	98.03	95.15	96.84	95.98	91.87
Genetic Advance (5%)	F_2	10.48	10.01	12.94	11.45	9.53	10.42
	F_3	8.26	11.01	7.58	13.55	10.26	7.29
Genetic Advance as % of Mean	F_2	60.10	56.86	67.94	53.63	46.39	64.22
	F_3	51.09	57.15	36.38	54.96	46.33	36.05

(Source: Narasimman, 2005)

9.3.3. Heterosis and inbreeding depression

Rice being a self pollinated crop, the scope for exploitation of hybrid vigour depends on the magnitude of heterosis, biological feasibility and the types of gene action involved. From the economic point of view, retention of heterosis in F_2 and later generations is of practical importance. Low inbreeding depression might be the reason for residual heterosis. The cost of hybrid seed production in rice is high. Hence, there is an urgent need to find an alternate technique to produce hybrid seeds at cheaper cost. Earlier reports indicated the scope of utilizing the second generation hybrid seed (F_2 seeds) in tomato (Larson and Currance, 1937), summer squash (Curtis, 1941), sesame (Ganesan, 1995) and bhendi (Senthil Kumar, 1998; Saravanan, 2001). Keeping this in view, the present inquiry was conducted to know whether there is any profitable retention of hybrid vigour in the F_2 and F_3 generations (residual heterosis) of rice for fifteen character are discussed here under.

The retention of heterosis may be due to the occurrence of transgressive segregants and close linkage of some favourable genes controlling these attributes. Further, this increased retention of the vigour may be due to the presence of additive genes derived from genetically divergent parental genotypes. In most of the cases, genetic diversity enhances the heterotic expression as well as helps for the retention of heterosis.

9.3.4. Scaling test and gene effects

The scaling test indicated that the scale C or D were significant for all the fifteen character, in all the six crosses studied. It is indicating the presence of epistatic interaction for all the traits of interest, in all the crosses (Table 77).

Table 77 : Scaling test and gene action for grain yield per plant in rice

	Scale		Gene effects					Type of epistasis
	C	D	$\hat{m}$	$[\hat{d}]$	$[\hat{h}]$	$[\hat{i}]$	$[\hat{l}]$	
$P_1 \times P_2$	-59.25 ± 1.80**	-33.82 ±1.07**	17.44 ±0.23**	-3.06 ± 0.31**	13.56 ± 0.82**	6.55 ±1.01**	33.91 ± 2.80**	Complementary recessive
$P_1 \times P_3$	-67.99 ±2.11**	-21.99 ±1.01**	19.05 ±0.28**	-4.86 ± 0.27**	8.23 ± 0.91**	-6.39 ±1.11**	61.33 ± 3.30**	Complementary recessive
$P_2 \times P_3$	-70.22 ±1.68**	-25.88 ±1.11**	20.56 ±0.21**	-1.80 ± 0.26**	8.45 ± 0.85**	1.96 ± 0.98*	59.13 ± 2.74**	Complementary recessive
$P_2 \times P_1$	-62.69 ±2.24**	-21.79 ±1.18**	17.61 ±0.21**	3.06 ± 0.31**	7.03 ± 0.99**	10.20 ±1.13**	54.53 ± 3.36**	Complementary recessive
$P_3 \times P_1$	-61.97 ±2.16**	-11.27 ±1.35**	21.35 ±0.24**	4.86 ± 0.27**	3.66 ± 1.10**	6.90 ±1.13**	67.60 ± 3.50**	Complementary recessive
$P_3 \times P_2$	-83.54 ±2.61**	-24.84 ±0.97**	16.22 ±0.23**	1.80 ± 0.26**	3.53 ± 1.02**	6.23 ± 1.16	78.27 ± 3.78**	Complementary recessive

* Significant at 5 per cent
** Significant at 1 per cent
(Source: Narasimman, 2005)

9.3.5. Gene Effects

The studies on gene effects in generation mean analysis revealed that additive gene effect was significant in all the crosses for days to first flowering, plant height at first flowering, plant height at maturity, productive tillers, boot leaf area, filled grains spikelet sterility, 100 grain weight, grain length, biomass per plant and grain yield per plant. For number of leaves at first flowering, the additive effect was significant only in $P_3 \times P_2$. The result indicated that there exist scope for direct selection for earliness, grain quality, seed yield and partitioning efficiency. Additive gene effect for plant height and productive tillers per plant was reported by Robin (1997). Dhanakodi and Subramanian (1998) reported additive gene effect for test grain weight.

The dominance gene effect was significant in all the crosses for days to first flowering, plant height at first flowering, plant height at maturity, total tillers per plant, spikelet sterility, 100 grain weight, grain length, grain L/B ratio, biomass per plant, grain yield per plant and harvest index. Dominance effect was significant in P_3P_2 for number of leaves at first flowering. In the remaining five crosses, it was significant for this character. Similarly, for productive tillers, dominance effect was non-significant in the cross $P_3 \times P_1$ whereas it was significant in all the remaining five crosses for this character. For boot leaf area, dominance effect was non-significant in $P_2 \times P_3$ whereas it was significant in the remaining five crosses. Dominance effect was non-significant in the cross $P_1 \times P_2$ for filled grains per panicle. In the remaining five crosses it was significant for this character. The result indicated the presence of dominance effect too, in the inheritance of the trail of interest. Partial dominance effect for days to 50 per cent flowering was reported by Agarwal and Sharma (1987). Dominance gene effect for productive tillers per plant was reported by Ram (1994) and Koodalingam (1994).

The additive × additive interaction effect was significant in all the crosses for days to first flowering, plant height at first flowering, plant height at maturity, total tillers per plant, boot leaf area and filled grains per panicle. The additive × additive interaction effect was non-significant in $P_3 \times P_1$ for number of leaves at first flowering. However, in all the other crosses it was significant for this character. Similarly, the cross $P_3 \times P_2$ showed non-significant additive × additive interaction effect for productive tillers per plant whereas the remaining five crosses showed significant effect. The additive × additive interaction effect for

100 grain weight was non-significant for 100 grain weight and gram length in the cross $P_3 \times P_2$. However, it was significant in all the other crosses for both the traits. For grain L/B ratio, excepting the cross $P_3 \times P_1$, the all other five crosses exhibited significant additive × additive interaction effect. Likewise, for biomass per plant, excepting $P_1 \times P_3$, all other crosses evinced significant additive × additive interaction effect. The cross $P_3 \times P_2$ exhibit non-significant additive × additive interaction effect for grain yield per plant. However, all the other crosses it was significant. For harvest index, the cross $P_1 \times P_2$ and its reciprocal even demonstrated non-significant additive × additive effect. The remaining four crosses showed significant additive × additive interaction effect.

The results are in conformity with the earlier reports of Guo and Wu (1988) for days to 50 per cent flowering; Subbaraman (1984), Dhanakodi (1990) and Vaithilingam (1995) for plant height; Bui Chi Buu and Phung Ba Tao (1992) and Vaithilingam (1995) for productive tiller number; Subbaraman (1984), Dhanakodi (1990), Nadarajan and Kumaravelu (1994), Koodalingam (1995), Vaithilingam (1995) and Robin (1997) for number of grains per panicle; Shaalan and Aly (1977), Ganapathy (1989), Dhanakodi (1990) and Vaithilingam (1991) for 1000 grain weight; and Ushakumari (1995), Vaithilingam (1995), Robin (1997) and Dhanakodi and Subramanian (1998) for grain yield.

The dominance × dominance interaction effect was significant in all the crosses for days to first flowering, plant height at first flowering, plant height at maturity, total tillers per plant, productive tillers per plant, boot leaf area, number of filled grains per panicle, spikelet sterility, 100 grain weight, grain L/B ratio, grain yield per plant and harvest index. For number of leaves at first flowering, this effect was non-significant with P_2P_1 and $P_3 \times P_1$. In the remaining four crosses it was significant for this character. For grain length, it was non-significant with $P_3 \times P_2$. In the remaining crosses, it was significant. The crosses $P_2 \times P_3$ had non-significant dominance × dominance interaction effect for biomass on plant. However, in the other five crosses this effect was significant.

Similar reports were reported for plant height (Subbaraman, 1984 and Vaithilingam, 1995); productive tillers (Koodalingam, 1994 and Vaithilingam 1995); grains per panicle (Subbaraman, 1994; Koodalingam, 1995; Nadarajan and Kumaravelu, 1994; Vaithilingam, 1995 and Robin, 1997); 1000 grain weight (Shaalan and Aly, 1977 and Ganapathy, 1989) and for grain yield (Ganapathy, 1989; Dhanakodi, 1990; Roy and Panwar, 1993 and Ushakumari, 1995).

The $[\hat{h}]$ and $[\hat{l}]$ effects took opposite sign for days to first flowering in five crosses. It indicated the presence of duplicate dominant epistasis in the inheritance of this trait. Only one cross (P_2P_1) exhibited complementary recessive epistasis for this character. For plant height at first flowering, the crosses viz., $P_1 \times P_2$, $P_1 \times P_3$ and $P_3 \times P_1$ demonstrated duplicate dominant epistasis. The remaining three crosses showed complementary recessive epistasis. For number of leaves at first flowering, all the crosses showed duplicate dominant epistasis. For plant height at maturity, the crosses viz., $P_1 \times P_2$, $P_1 \times P_3$, $P_3 \times P_1$ and $P_3 \times P_3$ evinced duplicate dominant epistasis. The other two crosses showed complementary recessive epistasis. Roy and Panwar (1993, 1997) reported duplicate epistasis for days to 50 per cent flowering. Dhanakodi (1990) reported duplicate dominant interaction for plant height while Vaithilingam (1995) reported duplicate epistasis.

Total tillers per plant, was controlled by duplicate dominant epistasis in all the crosses. For productive tillers per plant the crosses viz., $P_1 \times P_2$, $P_1 \times P_3$, P_2P_1 and $P_3 \times P_2$ showed duplicate dominant epistasis. The other two crosses showed complementary recessive epistasis. For boot leaf area, the crosses $P_1 \times P_2$, $P_2 \times P_3$ and $P_3 \times P_1$ demonstrated the presence of complementary recessive epistasis. The other three crosses showed duplicate dominant epistasis. For filled grains per panicle, all the crosses witnessed duplicate dominant epistasis. Vaithilingam (1995) reported duplicate epistasis for number of productive tillers. For grains per panicle, Koodalingam (1994) reported duplicate epistasis.

Spikelet sterility was controlled by duplicate dominant epistasis in three crosses viz., $P_1 \times P_2$, $P_1 \times P_3$ and $P_3 \times P_1$. In the remaining three crosses, it was controlled by complementary recessive epistasis. For 100 grain weight, five out of the six crosses showed the presence of duplicate dominant epistasis. In one cross ($P_3 \times P_2$), it was controlled by complementary recessive epistasis. Grain length was controlled by duplicate dominant epistasis in all the crosses. Grain L/B ratio, was controlled by duplicate dominant epistasis in five out of the six crosses. In the cross $P_2 \times P_1$, it was controlled by complementary recessive epistasis. For test grain weight, Dhanakodi (1990) observed duplicate dominance system.

Biomass per plant was controlled by complementary recessive epistasis in five out of the six crosses. In one cross viz., $P_3 \times P_1$ it was controlled by duplicate dominant epistasis. Grain yield per plant was controlled by complementary recessive epistasis in all the crosses. For harvest index, presence of complementary recessive epistasis was evidenced in four out of the six crosses. In two crosses viz., $P_2 \times P_3$ and $P_3 \times P_2$, the presence of duplicate dominant epistasis was also identified. For grain yield, Ganapathy (1989) reported duplicate dominance epistasis and Chauhan *et al.* (1993) reported duplicate epistasis.

It could be noted that the presence of additive, dominance, additive × additive interaction effects and dominance × dominance interaction effects were present along with either duplicate dominant epistasis or complementary recessive epistasis for most of the earliness, quality, yield and partitioning efficiency characters. Hence, selection in the early segregating generations may not give desirable recombinants. This may possibly be overcome by delaying the selection to later segregating generations when the dominance and epistasis disappear and resorting to intermating of segregants followed by recurrent selection Anderson (1939), Al-Jibouri *et al.*, (1958), Jensen (1970), Hallauer (1981 and 1986), Ramage (1981), Frey (1984) and Delogu *et al.* (1988) suggest recurrent selection as a basic breeding approach in autoganous crops. Diallel selective mating design suggested by Jensen (1970) can also be adopted, which will promote more recombination.

Chapter 10

Genetic Analysis of Back Cross F_1 Populations

The relative magnitude of different gene effects and an understanding of the mode of inheritance of complex quantitative traits have a direct bearing on the method of hybridization and selection which should be adopted in a specific breeding programme. Useful biometrical methods have been developed by the Birmingham group of scientists, to analyse the genetic architecture in organisms, through the use of first and second degree statistics obtainable from the true breeding inbreds, their first generation crosses and the different types of segregating progenies (Mather, 1949; Hayman, 1958; Gamble, 1962; Mather and Jinks, 1971). Since then this method has been widely used in the genetic analysis of various crops.

10.1. Generation mean analysis

Scaling test

The means of the different generations were utilized for obtaining the various genetic effects. The data were first tested to fit in with simple additive-dominance model without interaction. The scaling tests A, B and C (Mather and Jinks, 1971) were applied to test the fitness of the data to a simple additive-dominance model. The A, B and C values are calculated by using the following formulae.

$$A = 2\bar{B}_1 - \bar{P}_1 - \bar{F}_1$$

$$B = 2\bar{B}_2 - \bar{P}_2 - \bar{F}_1$$

$$C = 4\bar{F}_2 - 2\bar{F}_1 - \bar{P}_1 - \bar{P}_2$$

where $\bar{P}_1$, $\bar{P}_2$, $\bar{F}_1$, $\bar{F}_2$, $\bar{B}_1$ and $\bar{B}_2$ are the means of parent 1, parent 2, F_1, F_2 and back cross generations B_1 and B_2 respectively.

The variances of the three scales were computed as follows:

$$V_A = 4V_{B_1} + V_{P_1} + V_{F_1}$$

$$V_B = 4V_{B_2} + V_{P_2} + V_{F_1}$$

$$V_C = 16V_{F_2} + 4V_{F_1} + V_{P_1} + V_{P_2}$$

where, V_{P_1}, V_{P_2}, V_{F_1}, V_{F_2}, V_{B_1} and V_{B_2} are the variances of the means of the respective generations. The standard errors of A, B and C were calculated by estimating the square root of the variance concerned.

S.E. of A = $(V_A)^{1/2}$

S.E. of B = $(V_B)^{1/2}$

S.E. of C = $(V_C)^{1/2}$

The significance of the scales A, B and C was determined by comparing the observed the expected 't' values

$$t \text{ for } A = \frac{A}{(V_A)^{1/2}}$$

$$t \text{ for } B = \frac{B}{(V_B)^{1/2}}$$

$$t \text{ for } C = \frac{C}{(V_C)^{1/2}}$$

10.2. Estimation of genetic effects m, [d] and [h] on the assumption of the simple additive-dominance model

Joint scaling test

A joint scaling test was conducted following the procedure proposed by Cavalli (1952). Data from the six generations *viz.,* P_1, P_2, F_1, F_2, B_1 and B_2 were utilised to estimate the mid parent m, additive [d] and dominance [h] genetic effects by the method of the least squares

(Mather and Jinks, 1971). The adequacy of the additive-dominance model without interactions was tested by the chi-square test of goodness of fit of the observed mean values with the means estimated from the above three parameters for three degrees of freedom.

10.3. Estimation of genetic effects m, [d], [h], [i], [j] and [l] on the assumption of the presence of digenic interaction

The inadequacy of the data to fit in with a simple additive-dominance model without interaction required the extension of analysis of estimating the parameters describing epistasis. Using the six generation means, the estimates of mid-parent m, additive effect [d], dominance effect [h], additive × interaction [i], additive × dominance interaction [j] and dominance × dominance interaction [l] were obtained by a perfect fit method from the equations formulated by Mather and Jinks (1971).

$$m = \frac{1}{2}\bar{P}_1 + \frac{1}{2}\bar{P}_2 + 4\bar{F}_2 - 2\bar{B}_1 - 2\bar{B}_2$$

$$[d] = \frac{1}{2}\bar{P}_1 - \frac{1}{2}\bar{P}_2$$

$$[h] = 6\bar{B}_1 + 6\bar{B}_2 - 8\bar{F}_2 - \bar{F}_1 - 1_{1/2}\bar{P}_1 - 1_{1/2}\bar{P}_2$$

$$[i] = 2\bar{B}_1 + 2\bar{B}_2 - 4\bar{F}_2$$

$$[j] = 2\bar{B}_1 - \bar{P}_1 - 2\bar{B}_2 + \bar{P}_2$$

$$[l] = \bar{P}_1 + \bar{P}_2 + 2\bar{F}_1 + 4\bar{F}_2 - 4\bar{B}_1 - 4\bar{B}_2$$

The variances for different parameters were calculated using the squared values of the standard errors of the respective generation means in the following equations.

$$V_m = \frac{1}{4}V_{P_1} + \frac{1}{4}V_{P_2} + 16V_{F_2} + 4V_{B_1} + 4V_{B_2}$$

$$V_{[d]} = \frac{1}{4}V_{P_1} + \frac{1}{4}V_{P_2}$$

$$V_{[h]} = 36V_{B_1} + 36V_{B_2} + 64V_{F_2} + V_{F_1} + 9/4V_{P_1} + 9/4V_{P_2}$$

$$V_{[i]} = 4V_{B_1} + 4V_{P_2} + 16V_{F_2}$$

$$V_{[j]} = 4V_{B_1} + V_{P_1} + 4V_{B_2} + V_{P_2}$$

$$V_{[l]} = V_{P_1} + V_{P_2} + 4V_{F_1} + 16V_{F_2} + 16V_{B_1} + 16V_{B_2}$$

The standard error was obtained by the square root of the respective variances. The significance of these parameters were tested by calculating the 't' value as done for scaling test.

10.4. Partitioning the variance components

The variance of the difference generations were partitioned into different genetic components of variation following Mather and Jinks (1977).

The estimates of four components of variation D, H, F and E_W were obtained as follows:

$$E_W = \frac{1}{4}\left(V_{P_1} + V_{P_2} + 2V_{F_1}\right)$$

$$D = 4V_{F_2} - 2\left(V_{B_1} - V_{B_2}\right)$$

$$H = 4\left(V_{B_1} + V_{B_2} - V_{F_2} - E_W\right)$$

$$F = V_{B_2} - V_{B_1}$$

where,

E_W = Non heritable variance due to environment,

D = Fixable variance due to additive genes.

H = Non fixable variance due to dominance and

F = The covariance of additive and dominance effects.

V_{P_1}, V_{P_2}, V_{F_1}, V_{B_1} and V_{B_2} are the variance of the respective generations.

Dominance ratio

The dominance ratio was worked out from the D and H estimates as follows:

Dominance ratio = $(H/D)^{1/2}$

10.5. Example

Thirugnanakumar (1991) made generation mean and variance analysis with three direct and three reciprocal crosses in sesame.

10.6. Summary of the results on gene effects

The scaling tests indicated the inadequacy of the data to a simple additive - dominance model for all the characters and crosses in general with a few exceptions. Simple additive-dominance model was adequate in the direct cross, $P_2 \times P_3$, for number of branches, density of seeds and harvest index. For its reciprocal cross it was inadequate for the same traits. The adequacy in both direct and reciprocal crosses ($P_2 \times P_3$, $P_3 \times P_2$) was found for volume of 1000 seeds. It was also adequate in the reciprocal crosses *viz.*, $P_2 \times P_1$ and $P_3 \times P_1$ for harvest index and inadequate in their direct crosses for the same trait. The lack of adequacy of this simple model in the reciprocal cross of $P_2 \times P_3$ i.e. in $P_3 \times P_2$ for number of branches, seed density and harvest index well as in the direct crosses of the reciprocal crosses viz., $P_2 \times P_1$ and $P_3 \times P_1$ i.e. in $P_1 \times P_2$ and $P_1 \times P_3$ indicated the existence of reciprocal differences for these traits.

The existence of simple additive-dominance model indicated the prevalence of dominance effect for these traits in the respective crosses. Devoiding these cross combinations, all other crosses were subjected to the additive-dominance model with digenic interaction. The overall genetic situation obtained for the sixteen traits of interest in different cross combination is summarized in Tables 198 and 199.

For days to maturity, additive, dominance and epistatic effects were significant in most of the crosses with predominance of dominance gene action. This was again conformed by the preponderance of dominance variance than the additive variance. The dominance ratio was also more than unity. Duplicate type of epistasis was present in all the crosses.

Dominance and epistatic effects were significant for plant height at maturity. dominance was predominant. Dominance variance was higher than additive variance. Duplicate dominant epistasis was present in all crosses.

On overall basis, additive, dominance and epistatic effects were present for number of branches with predominance of non-additive epistatic effect. H component was higher in magnitude than D component and the dominance ratio was more than unity. Dominant duplicate epistasis was observed in all croses.

Dominance × dominance epistatic effect was preponderant for number of flowers. D was higher than H in four crosses and the reverse was true in two crosses. The dominance ratio was more than unity in two crosses. Duplicate dominant epistasis was present in all the crosses.

For number of capsules per plant, additive, dominance and epostatic effects were significant. Off these, dominance and additive × additive epistatic effects were important. Estimates of H was higher in magnitude in five crosses and the dominance ratio was more than unity in two crosses. Estimates of D was higher than H in two crosses and dominance ratio was less than unity in two crosses, conforming the prevalence of both dominance and additive × additive epistatic effects. Five crosses showed duplicate dominant epistasis and one cross showed complementary recessive epistasis.

Additive, dominance and epistatic effects were significant for volume of capsules. However, dominance and dominance × dominance epistatic effects were preponderant. H was higher than D in five crosses. In one cross Estimates of D was more than H component. Dominance Ratio was less than unity in one cross. Dominant duplicate epistasis was present in five crosses. In one cross, recessive epistasis was present.

Dominance followed dominance × dominance interaction effect were preponderant even though all the three effects were significant for number of seed per capsule. H was higher than D in all the crosses. The dominance ratio was more than unity in two crosses. Duplicate dominant epistasis was present in all the crosses.

For seed length, additive, dominance and epistatic effects were significant with predominance of dominance and dominance × dominance interaction effect. However, D was higher than H in four crosses and H was higher than D in two crosses. Dominance ratio was less than unity in two crosses indicating the importance of additivity too. Duplicate dominant epistasis was present in all the crosses.

Additive, dominance and epistatic effects were significant. For seed breadth, dominance was higher in magnitude. H was more than D in five crosses. D was more than H in only one cross. Dominance ratio was more than unity in one cross. Duplicate dominant epistasis was present in all the crosses.

Additive, dominance and epistatic effects were prevalent for weight of 1000 seeds. Dominance was higher in magnitude. H was more than D in five crosses and H was equal to D in one cross. Dominance ratio was more than unity in one cross and unity in one cross indicating the presence of additivity too. Duplicate dominant epistasis was present in all the crosses.

For 1000 seed volume all the effects were non-significant. H was higher than D in four crosses and D was higher than H in two crosses, indicating the importance of both dominance and additivity. Four crosses showed duplicate dominant epistasis.

Dominance × dominance interaction effect was important in the expression of seed density even though additivity, and dominance were also observed. Dominance was higher than additive variance in all the crosses. Dominance ratio was more than unity in two crosses. Five crosses showed duplicate dominant epistasis.

For oil content, additive, dominance and epistatic effects were significant. Dominance was higher in magnitude. Dominance was higher than additive variance in all the crosses. Dominance ratio was more than unity in two crosses. In all the crosses duplicate dominant epistasis was present.

For seed yield, dominance and epistatic effects were significant. Dominance was higher in magnitude. Additive × additive interaction effect was alone significant in two crosses. Dominance variance was higher than additivity in all the crosses. The dominance ratio was more than unity in one cross. Duplicate dominant epistasis was present in five crosses and recessive complementary epistasis was present in one cross.

Additive, dominance and epistatic interaction effects were significant for total dry matter production. Dominance and dominance × dominance interaction effect was prevalent. In one cross, additive and additive epistasis was alone significant. H was higher than D in all the crosses. Dominance ratio was more than unity in three crosses. Duplicate dominant epistasis was present in all the crosses.

For harvest index, dominance × dominance interaction effect alone was significant that too in only one cross. Other crosses showed non-significant genetic effects. D was higher than H in all the crosses. Both duplicate and dominant (two crosses) and complementary recessive epistasis was present.

Table 78 : Scaling and joint scaling tests for days to maturity

Cross combinations	Scales A	B	C	χ^2 (3)	P
$P_1 \times P_2$	21.51** ± 1.67	6.70** ± 2.03	5.13 ± 3.31	174.49**	< 0.001
$P_1 \times P_3$	13.39** ± 2.18	10.26** ± 2.01	18.57** ± 2.90	92.53**	< 0.001
$P_2 \times P_3$	-4.90* ± 1.96	-5.20 ± 2.95	4.32 ± 3.19	11.89**	0.01-0.001
$P_2 \times P_1$	-9.60** ± 1.94	0.95 ± 1.98	5.25 ± 3.06	30.65**	< 0.001
$P_3 \times P_1$	5.31* ± 2.06	5.20** ± 1.63	4.55** ± 2.50	267.79**	< 0.001
$P_3 \times P_2$	4.60** ± 1.72	-2.96** ± 1.44	223.24 ± 6.79	57.60**	< 0.001

(Source: Thirugnanakumar, 1991)

Table 79 : Scaling and joint scaling test for height of the plant at maturity (cm)

Cross combinations	Scales A	B	C	χ^2 (3)	P
$P_1 \times P_2$	58.22** ± 4.93	49.35** ± 3.57	49.07** ± 7.10	315.29**	< 0.001
$P_1 \times P_3$	36.94** ± 4.07	41.35** ± 5.28	52.95** ± 7.69	160.65**	< 0.001
$P_2 \times P_3$	25.84** ± 4.33	5.04 ± 3.98	47.04** ± 7.03	70.17**	< 0.001
$P_2 \times P_1$	5.46 ± 5.08	-2.09 ± 3.26	40.49** ± 7.81	29.73**	< 0.001
$P_3 \times P_1$	9.15** ± 3.31	13.48** ± 3.86	75.07** ± 8.58	84.22**	< 0.001
$P_3 \times P_2$	9.84** ± 3.95	2.68 ± 3.20	-9.56 ± 6.34	10.85*	0.02-0.01

(Source: Thirugnanakumar, 1991)

Table 80 : Scaling and joint scaling test for number of branches

Cross combinations	Scales A	B	C	χ^2 (3)	P
$P_1 \times P_2$	4.31** ± 0.75	10.06 ± 0.91	-1.25 ± 1.27	155.67**	< 0.001
$P_1 \times P_3$	4.17** ± 1.03	1.26 ± 0.64	4.95** ± 1.39	34.30**	< 0.001
$P_2 \times P_3$	-0.36 ± 0.64	-0.80 ± 0.72	-1.52 ± 1 .17	02.58	< 0.001
$P_2 \times P_1$	-2.86** ± 0.69	10.00 ± 0.66	-2.97* ± 1.41	19.81**	< 0.001
$P_3 \times P_1$	-0.21 ± 0.76	1.84* ± 0.72	4.45** ± 1.26	17.76**	< 0.001
$P_3 \times P_2$	0.99 ± 0.62	1.79** ± 0.66	1.22 ± 1.11	10.11*	0.02-0.01

* Significant at P = 5%; ** Significant at P = 1%
(Source: Thirugnanakumar, 1991)

Table 81 : Scaling and joint scaling tests for number of flowers

Cross combinations	Scales A	B	C	χ^2 (3)	P
$P_1 \times P_2$	103.59** ± 12.26	53.24** ± 9.82	73.77** ± 9.75	139.60**	< 0.001
$P_1 \times P_3$	62.39** ± 9.05	42.79** ± 4.63	117.32** ± 14.08	173.09**	< 0.001
$P_2 \times P_3$	27.70** ± 6.21	38.25** ± 4.92	87.03** ± 15.53	86.06**	< 0.001
$P_2 \times P_1$	33.75** ± 5.36	28.80** ± 6.04	68.75** ± 11.22	72.78**	< 0.001
$P_3 \times P_1$	35.34** ± 6.75	42.54** ± 5.24	97.22** ± 17.64	91.26**	< 0.001
$P_3 \times P_2$	58.91** ± 6.14	39.02** ± 4.83	68.19** ± 10.79	153.66**	< 0.001

(Source: Thirugnanakumar, 1991)

Table 82 : Scaling and joint scaling test for number of capsules

Cross combinations	Scales A	B	C	χ^2 (3)	P
$P_1 \times P_2$	98.35** ± 11.35	24.20** ± 9.14	35.35** ± 12.75	86.85**	< 0.001
$P_1 \times P_3$	57.40** ± 9.23	31.10** ± 5.22	75.90** ± 15.53	90.50**	< 0.001
$P_2 \times P_3$	11.55* ± 5.61	17.40** ± 6.05	74.35** ± 15.05	34.12**	< 0.001
$P_2 \times P_1$	3.50 ± 6.73	16.45** ± 5.22	57.55** ± 12.90	27.98**	< 0.001
$P_3 \times P_1$	33.10** ± 6.69	22.80** ± 5.22	97.50** ± 14.75	78.81**	< 0.001
$P_3 \times P_2$	38.30** ± 6.14	2.35 ± 3.97	24.95** ± 8.91	44.06**	< 0.001

(Source: Thirugnanakumar, 1991)

Table 83 : Scaling and joint scaling test for volume of capsules (cm^3)

Cross combinations	Scales A	B	C	χ^2 (3)	P
$P_1 \times P_2$	0.19** ± 0.03	0.66** ± 0.04	0.21** ± 0.06	252.61**	< 0.001
$P_1 \times P_3$	0.26** ± 0.04	0.50 ± 0.04	0.32** ± 0.07	173.88**	< 0.001
$P_2 \times P_3$	0.67** ± 0.05	0.40** ± 0.04	0.75** ± 0.08	304.99**	< 0.001
$P_2 \times P_1$	0.55** ± 0.04	0.48** ± 0.04	0.19** ± 0.06	315.93**	< 0.001
$P_3 \times P_1$	0.09** ± 0.03	0.23** ± 0.04	0.02 ± 0.07	46.53**	< 0.001
$P_3 \times P_2$	0.19** ± 0.03	0.42** ± 0.03	1.01** ± 0.08	320.73**	< 0.001

* Significant at P = 5%; ** Significant at P = 1%
(Source: Thirugnanakumar, 1991)

Table 84 : Scaling and joint scaling tests for number of seeds per capsule

Cross combinations	Scales A	B	C	χ^2 (3)	P
$P_1 \times P_2$	1.80* ± 0.77	3.60** ± 0.89	3.60** ± 1.38	21.43**	< 0.001
$P_1 \times P_3$	-0.80 ± 0.79	4.79 ± 5.09	3.20* ± 1.51	32.49**	< 0.001
$P_2 \times P_3$	5.60** ± 0.83	4.40** ± 0.91	9.20** ± 1.34	79.47**	< 0.001
$P_2 \times P_1$	4.80** ± 0.84	2.80 ± 2.51	3.20* ± 1.44	34.22**	< 0.001
$P_3 \times P_1$	2.80** ± 2.53	2.80 ± 0.84	-1.60 ± 1.53	22.48**	< 0.001
$P_3 \times P_2$	1.60* ± 0.83	5.80** ± 0.84	8.80** ± 1.61	63.02**	< 0.001

(Source: Thirugnanakumar, 1991)

Table 85 : Scaling and joint scaling test for length of seeds (mm)

Cross combinations	Scales A	B	C	χ^2 (3)	P
$P_1 \times P_2$	-0.06 ± 0.05	-0.23*8 ± 0.04	-0.67** ± 0.14	59.93**	< 0.001
$P_1 \times P_3$	-0.04 ± 0.05	0.15 ± 0.08	-0.55** ± 0.10	37.48**	< 0.001
$P_2 \times P_3$	-0.19** ± 0.04	0.11** ± 0.04	-0.02 ± 0.10	33.98**	< 0.001
$P_2 \times P_1$	-0.44** ± 0.05	-0.61** ± 0.06	-0.77** ± 0.11	200.48**	< 0.001
$P_3 \times P_1$	-0.09* ± 0.04	-0.26** ± 0.04	-0.01 ± 0.10	39.48**	< 0.001
$P_3 \times P_2$	0.04 ± 0.04	-0.16** ± 0.05	0.28** ± 0.11	23.24**	< 0.001

(Source: Thirugnanakumar, 1991)

Table 86 : Scaling and joint scaling test for breadth of seeds (mm)

Cross combinations	Scales A	B	C	χ^2 (3)	P
$P_1 \times P_2$	-0.12** ± 0.03	-0.21** ± 0.03	-0.29** ± 0.06	64.64**	< 0.001
$P_1 \times P_3$	-0.06 ± 0.04	-0.09** ± 0.03	-0.33** ± 0.07	27.31**	< 0.001
$P_2 \times P_3$	-0.04 ± 0.06	0.06 ± 0.05	0.16** ± 0.06	12.39**	< 0.001
$P_2 \times P_1$	-0.17** ± 0.03	-0.28* ± 0.03	-0.05 ± 0.07	78.67**	< 0.00
$P_3 \times P_1$	-0.06 ± 0.04	-0.21** ± 0.04	0.29** ± 0.08	65.35**	< 0.001
$P_3 \times P_2$	-0.03 ± 0.05	0.09* ± 0.04	0.22** ± 0.07	13.42**	< 0.001

* Significant at P = 5%; ** Significant at P = 1%
(Source: Thirugnanakumar, 1991)

Table 87 : Scaling and joint scaling tests for weight of 1000 seeds (g)

Cross combinations	Scales A	B	C	χ^2 (3)	P
$P_1 \times P_2$	1.42** ± 0.26	1.08** ± 0.26	0.86 ± 0.52	49.63**	< 0.001
$P_1 \times P_3$	0.30 ± 0.23	2.06** ± 0.33	2.48** ± 0.55	60.33**	< 0.001
$P_2 \times P_3$	1.62** ± 0.32	1.06* ± 0.22	1.94** ± 0.51	62.68**	< 0.001
$P_2 \times P_1$	2.42** ± 0.43	0.82** ± 0.35	2.06** ± 0.63	48.22**	< 0.001
$P_3 \times P_1$	1.49** ± 0.31	-0.33 ± 0.24	1.10* ± 0.51	30.38**	< 0.001
$P_3 \times P_2$	1.38** ± 0.31	-0.18 ± 0.13	2.66** ± 0.57	44.00**	< 0.001

(Source: Thirugnanakumar, 1991)

Table 88 : Scaling and joint scaling test for volume of 1000 seeds (ml)

Cross combinations	Scales A	B	C	χ^2 (3)	P
$P_1 \times P_2$	4.41 ± 3.00	8.33* ± 3.27	12.58* ± 5.27	11.39**	0.01-0.001
$P_1 \times P_3$	-0.19 ± 2.14	0.10** ± 2.09	6.85 ± 5.97	1.46	0.95-0.50
$P_2 \times P_3$	2.47 ± 2.17	7.58 ± 4.27	7.97 ± 5.60	5.05	0.20-0.10
$P_2 \times P_1$	5.10* ± 2.39	3.40 ± 2.43	4.84 ± 6.28	5.63	0.20-0.10
$P_3 \times P_1$	1.03 ± 2.00	6.00* ± 2.98	6.79 ± 5.72	4.78	0.20-0.10
$P_3 \times P_2$	-2.18 ± 1.97	0.87 ± 2.16	-1.43 ± 4.56	1.87	0.95-0.50

(Source: Thirugnanakumar, 1991)

Table 89 : Scaling and joint scaling test for density of seeds (g/ml)

Cross combinations	Scales A	B	C	χ^2 (3)	P
$P_1 \times P_2$	1.83** ± 0.37	1.48* ± 0.59	0.83 ± 0.48	32.21**	< 0.001
$P_1 \times P_3$	0.75** ± 0.25	1.11** ± 0.29	1.91** ± 0.53	32.50**	< 0.001
$P_2 \times P_3$	1.97 ± 0.42	0.74 ± 0.19	1.06 ± 0.41	37.94**	< 0.001
$P_2 \times P_1$	0.60* ± 0.25	0.74** ± 0.27	1.33* ± 0.53	17.52**	< 0.001
$P_3 \times P_1$	0.88** ± 0.39	2.32** ± 0.49	2.42* ± 0.93	32.89**	< 0.001
$P_3 \times P_2$	2.11* ± 0.38	1.38** ± 0.29	1.39** ± 0.37	60.24**	< 0.001

* Significant at P = 5%; ** Significant at P = 1%

(Source: Thirugnanakumar, 1991)

Table 90 : Scaling and joint scaling tests for oil content (%)

Cross combinations	Scales A	B	C	χ^2 (3)	P
$P_1 \times P_2$	2.89** ± 1.09	2.71 ± 1.69	12.40** ± 2.38	35.53**	< 0.001
$P_1 \times P_3$	-3.24* ± 1.50	1.47 ± 1.39	6.79** ± 2.48	13.38**	0.01-0.001
$P_2 \times P_3$	4.03** ± 1.27	-3.44** ± 1.13	10.55** ± 2.30	40.42**	< 0.001
$P_2 \times P_1$	0.56 ± 1.76	-2.38 ± 1.71	6.74** ± 2.30	10.84*	< 0.001
$P_3 \times P_1$	8.86** ± 1.09	6.13** ± 1.04	14.89** ± 2.31	128.53**	< 0.001
$P_3 \times P_2$	3.62** ± 1.39	7.33** ± 1.22	8.95** ± 2.85	48.86**	< 0.001

Table 91 : Scaling and joint scaling test for seed yield (g)

Cross combinations	Scales A	B	C	χ^2 (3)	P
$P_1 \times P_2$	15.28** ± 1.73	5.07** ± 1.22	6.47** ± 1.9i1	98.50**	< 0.001
$P_1 \times P_3$	6.16** ± 0.98	3.86** ± 0.77	8.86** ± 1.73	75.32**	< 0.001
$P_2 \times P_3$	1.69* ± 0.84	0.86 ± 0.80	5.81** ± 1.42	17.85**	< 0.001
$P_2 \times P_1$	1.65* ± 0.83	2.36** ± 0.87	6.59** ± 1.37	27.62**	< 0.001
$P_3 \times P_1$	1.04 ± 0.77	2.14** ± 0.76	6.42** ± 1.36	27.07**	< 0.001
$P_3 \times P_2$	3.37** ± 0.81	1.76* ± 0.82	3.07* ± 1.25	20.96**	< 0.001

Table 92 : Scaling and joint scaling test for total dry matter production (g)

Cross combinations	Scales A	B	C	χ^2 (3)	P
$P_1 \times P_2$	55.65** ± 4.16	23.12** ± 4.47	29.09** ± 8.14	206.63**	< 0.001
$P_1 \times P_3$	33.23** ± 3.71	23.12** ± 3.67	47.93** ± 7.56	149.50**	< 0.001
$P_2 \times P_3$	4.67 ± 3.20	-0.03 ± 2.93	32.06** ± 6.60	24.94**	< 0.001
$P_2 \times P_1$	3.95 ± 3.54	10.52** ± 3.90	35.87** ± 6.49	36.21**	< 0.001
$P_3 \times P_1$	4.02 ± 3.18	10.09** ± 3.27	40.45** ± 7.20	39.86**	< 0.001
$P_3 \times P_2$	17.49** ± 3.76	4.59 ± 2.71	15.06* ± 6.30	27.63**	< 0.001

Table 93 : Scaling and joint scaling tests for harvest index (%)

Cross combinations	Scales A	B	C	χ^2 (3)	P
$P_1 \times P_2$	13.39** ± 3.44	6.76* ± 3.38	15.81* ± 6.24	19.46**	< 0.01
$P_1 \times P_3$	4.12 ± 3.43	4.35 ± 3.47	13.19* ± 6.14	5.54	0.20-0.10
$P_2 \times P_3$	5.97 ± 3.63	5.73 ± 3.34	11.14 ± 6.27	6.07	0.20-0.10
$P_2 \times P_1$	5.66 ± 3.33	5.89 ± 3.33	10.29 ± 6.17	6.28	0.10-0.05
$P_3 \times P_1$	4.86 ± 3.18	4.33 ± 3.43	4.69 ± 6.34	3.37	0.50-0.30
$P_3 \times P_2$	7.49* ± 3.49	6.29 ± 3.47	9.70 ± 6.36	7.32	0.10-0.05

* Significant at P = 5%; ** Significant at P = 1%

(Source: Thirugnanakumar, 1991)

Table 94 : Genetic effects for days to maturity

Cross combinations	Parameters m	d	H	I	j	l
$P_1 \times P_2$	66.56** ± 4.01	4.22** ± 0.39	64.59** ± 9.77	23.08** ± 3.99	14.81** ± 2.58	-51.29** ± 5.93
$P_1 \times P_3$	80.15** ± 3.93	0.17 ± 0.42	30.73** ± 10.11	5.08 ± 3.91	3.13 ± 2.94	-28.73** ± 6.33
$P_2 \times P_3$	103.87** ± 4.56	4.40** ± 0.39	-45.84** ± 11.98	-14.42** ± 4.57	0.30 ± 3.51	24.52** ± 7.55
$P_2 \times P_1$	103.53** ± 3.78	4.22** ± 0.39	-37.82** ± 9.52	-13.90** ± 3.75	-10.55** ± 2.67	22.55** ± 5.96
$P_3 \times P_1$	115.27** ± 3.43	-0.17 ± 0.42	-53.70** ± 8.88	-30.04** ± 3.41	0.11 ± 2.62	19.53** ± 5.56
$P_3 \times P_2$	109.81** ± 3.75	-4.40** ± 0.39	46.58** ± 8.87	-20.36** ± 3.73	7.56 ± 2.21	18.72** ± 5.26

(Source: Thirugnanakumar, 1991)

Table 95 : Genetic effects for height of the plant at maturity (cm)

Cross combinations	Parameters m	d	H	I	j	l
$P_1 \times P_2$	-21.04* ± 8.55	-0.20 ± 0.85	221.75** ± 21.20	58.50** ± 8.51	8.87 ± 5.83	-166.07** ± 13.22
$P_1 \times P_3$	11.62 ± 9.50	-0.51 ± 0.85	123.99** ± 23.66	25.34** ± 9.46	-4.41 ± 6.45	-103.63** ± 14.64
$P_2 \times P_3$	53.33** ± 8.35	-0.30 ± 0.84	0.04 ± 20.67	-16.16 ± 8.31	20.80** ± 5.61	-14.72 ± 12.80
$P_2 \times P_1$	73.79** ± 9.14	0.20 ± 0.85	-64.67** ± 22.18	-37.12** ± 9.10	7.55 ± 5.79	33.75* ± 13.54
$P_3 \times P_1$	89.41** ± 9.30	0.51 ± 0.85	-78.89** ± 21.27	-52.44** ± 9.26	-4.33 ± 4.82	29.81* ± 12.44
$P_3 \times P_2$	15.09* ± 7.30	0.30 ± 0.84	68.20** ± 17.90	22.08** ± 7.25	7.16 ± 4.82	-34.60** ± 11.05

(Source: Thirugnanakumar, 1991)

Table 96 : Genetic effects for number of branches

Cross combinations	Parameters m	d	H	I	j	l
$P_1 \times P_2$	-10.00** ± 1.63	-0.47** ± 0.14	44.99** ± 4.13	15.62** ± 1.63	-5.75** ± 1.15	-29.99** ± 2.57
$P_1 \times P_3$	4.31* ± 1.77	-0.18 ± 0.14	6.38 ± 4.42	1.02 ± 1.77	3.45** ± 1.20	-6.99* ± 2.71
$P_2 \times P_3$	5.73** ± 0.13	0.31* ± 0.13	-0.52* ± 0.26	-	-	-
$P_2 \times P_1$	5.51** ± 1.56	0.47** ± 0.14	-1.54 ± 3.68	0.12 ± 1.56	-2.87** ± 0.90	2.73 ± 2.21
$P_3 \times P_1$	8.15** ± 1.52	0.18 ± 0.14	-4.39 ± 3.76	-2.82 ± 1.52	-2.05* ± 1.01	1.19 ± 2.32
$P_3 \times P_2$	4.24** ± 1.36	-0.30* ± 0.14	4.15 ± 8.30	1.56 ± 1.34	-0.80 ± 0.88	-4.34* ± 2.01

(Source: Thirugnanakumar, 1991)

Table 97 : Genetic effects for number of flowers

Cross combinations	Parameters m	d	H	I	j	l
$P_1 \times P_2$	-35.04* ± 17.57	-2.43 ± 1.63	318.08** ± 49.06	83.06** ± 17.49	50.35** ± 15.56	-239.89** ± 31.95
$P_1 \times P_3$	57.99** ± 16.56	-1.05 ± 1.37	74.60 ± 39.55	-12.14 ± 16.51	19.60* ± 9.93	-93.04* ± 23.72
$P_2 \times P_3$	69.36** ± 16.02	1.38 ± 1.59	21.07 ± 35.48	-21.00 ± 15.95	-10.55 ± 7.28	-44.87* ± 20.31
$P_2 \times P_1$	53.43** ± 12.16	2.43 ± 1.63	56.53 ± 28.99	-6.20 ± 12.05	4.95 ± 7.55	-56.35** ± 17.64
$P_3 \times P_1$	65.19** ± 18.06	1.05 ± 1.37	39.95 ± 39.66	-19.34 ± 18.00	-7.20 ± 7.61	-58.54** ± 22.65
$P_3 \times P_2$	18.53 ± 11.86	-1.38 ± 1.59	146.27** ± 28.30	29.74* ± 11.75	19.89** ± 7.39	-127.67** ± 17.16

* Significant at P = 5%; ** Significant at P = 1%

(Source: Thirugnanakumar, 1991)

Table 98 : Genetic effects for number of capsules

Cross combinations	Parameters m	d	H	I	j	l
$P_1 \times P_2$	-58.38** ± 18.93	-10.88** ± 0.90	292.63** ± 49.67	87.20** ± 18.91	74.15** ± 14.45	-209.75** ± 31.38
$P_1 \times P_3$	07.00 ± 18.39	-01.65 ± 0.89	114.55** ± 43.46	12.60 ± 18.37	26.30* ± 10.45	-101.10** ± 25.79
$P_2 \times P_3$	75.88** ± 16.72	9.23** ± 0.90	-66.98 ± 37.85	-45.04** ± 16.70	-5.85 ± 8.06	16.45 ± 21.76
$P_2 \times P_1$	66.42 ± 14.95	10.88** ± 0.90	-57.47 ± 35.06	-37.60* ± 14.92	-12.95 ± 8.32	17.65 ± 20.75
$P_3 \times P_1$	-61.20** ± 16.54	1.65 ± 0.89	-23.25 ± 37.78	-41.60* ± 16.52	10.30 ± 8.28	-14.30 ± 21.89
$P_3 \times P_2$	14.78 ± 10.85	-9.23** ± 0.90	61.83* ± 26.69	15.70 ± 10.82	35.95** ± 7.09	-56.35** ± 16.36

(Source: Thirugnanakumar, 1991)

Table 99 : Genetic effects for volume of capsules (cm^2)

Cross combinations	Parameters m	d	H	I	j	l
$P_1 \times P_2$	-0.44** ± 0.08	0.04** ± 0.01	2.10** ± 0.19	0.64** ± 0.08	-0.47** ± 0.05	-1.49** ± 0.12
$P_1 \times P_3$	-0.24** ± 0.08	0.04** ± 0.01	1.60** ± 0.20	0.44** ± 0.08	-0.24** ± 0.05	-1.20** ± 0.12
$P_2 \times P_3$	-0.16 ± 0.09	0.01 ± 0.01	1.73** ± 0.21	0.32** ± 0.09	0.27** ± 0.05	-1.39** ± 0.13
$P_2 \times P_1$	-0.64** ± 0.07	-0.04** ± 0.01	2.69** ± 0.19	0.84** ± 0.07	0.07 ± 0.05	-1.87** ± 0.12
$P_3 \times P_1$	-0.10 ± 0.08	-0.04** ± 0.01	0.99** ± 0.18	0.30** ± 0.08	-0.14** ± 0.05	-0.62** ± 0.11
$P_3 \times P_2$	0.57** ± 0.09	-0.01 ± 0.01	-0.19 ± 0.20	-0.40** ± 0.09	-0.23** ± 0.05	-0.21 ± 0.12

(Source: Thirugnanakumar, 1991)

Table 100 : Genetic effects for seeds per capsule

Cross combinations	Parameters m	d	H	I	j	l
$P_1 \times P_2$	48.00** ± 1.56	1.60** ± 0.19	7.80* ± 3.91	1.80 ± 1.55	-1.80 ± 1.09	-7.20** ± 2.47
$P_1 \times P_3$	50.00** ± 1.50	1.40** ± 0.23	2.20 ± 3.62	0.00 ± 1.40	-4.80** ± 0.97	-3.20 ± 2.28
$P_2 \times P_3$	47.60** ± 1.52	-0.20 ± 0.19	12.00** ± 3.88	0.80 ± 1.51	1.20 ± 1.12	-10.80** ± 2.50
$P_2 \times P_1$	46.20** ± 1.51	-1.60** ± 0.19	13.20** ± 3.74	3.60* ± 1.50	2.80** ± 1.02	-10.40** ± 2.39
$P_3 \times P_1$	43.60** ± 1.56	-1.40** ± 0.23	16.60** ± 3.77	6.40** ± 1.54	-0.80 ± 1.01	-11.20** ± 2.37
$P_3 \times P_2$	49.80** ± 1.69	0.20 ± 0.19	5.20 ± 4.06	-1.40 ± 1.68	-4.20** ± 1.04	-6.00* ± 2.53

(Source: Thirugnanakumar, 1991)

Table 101 : Genetic effects for volume of capsules (cm^2)

Cross combinations	Parameters m	d	H	I	j	l
$P_1 \times P_2$	2.65** ± 0.15	0.02 ± 0.01	0.26 ± 0.32	0.38** ± 0.25	0.17** ± 0.06	-0.09 ± 0.17
$P_1 \times P_3$	2.30** ± 0.13	0.11** ± 0.01	1.22** ± 0.33	0.66** ± 0.13	-0.19* ± 0.10	-0.77** ± 0.21
$P_2 \times P_3$	2.99** ± 0.11	0.08** ± 0.01	-0.47 ± 0.24	-0.06 ± 0.11	-0.30** ± 0.05	0.14 ± 0.14
$P_2 \times P_1$	3.32** ± 0.12	-0.02 ± 0.01	-1.69** ± 0.29	-0.28* ± 0.12	0.17* ± 0.07	1.33** ± 0.17
$P_3 \times P_1$	3.32** ± 0.10	-0.11** ± 0.01	-1.29** ± 0.23	-0.36** ± 0.10	0.17** ± 0.05	0.71** ± 0.14
$P_3 \times P_2$	3.33** ± 0.11	-0.08** ± 0.01	-1.16** ± 0.26	-0.40** ± 0.11	0.20** ± 0.06	0.52** 0.15

* Significant at P = 5%; ** Significant at P = 1%

(Source: Thirugnanakumar, 1991)

Table 102 : Genetic effects for breadth of seeds (mm)

Cross combinations	Parameters m	d	H	I	j	l
$P_1 \times P_2$	1.74** ± 0.07	0.02* ± 0.01	-0.52** ± 0.16	-0.04 ± 0.07	0.09* ± 0.04	0.37** ± 0.10
$P_1 \times P_3$	1.50** ± 0.09	0.04** ± 0.01	0.06* ± 0.20	0.18* ± 0.09	0.03 ± 0.05	-0.03 ± 0.12
$P_2 \times P_3$	1.80** ± 0.07	0.02** ± 0.01	-0.46* ± 0.20	-0.14* ± 0.07	-0.10 ± 0.06	0.12 ± 0.14
$P_2 \times P_1$	2.10** ± 0.08	-0.02* ± 0.01	-1.51** ± 0.18	-0.40** ± 0.08	0.11** ± 0.04	0.85** ± 0.11
$P_3 \times P_1$	2.24** ± 0.08	-0.04** ± 0.01	-1.55** ± 0.18	-0.56** ± 0.08	0.15** ± 0.04	0.83** ± 0.11
$P_3 \times P_2$	1.82** ± 0.08	-0.02** ± 0.01	-0.47* ± 0.21	-0.16 ± 0.08	-0.12** ± 0.06	0.10 ± 0.13

(Source: Thirugnanakumar, 1991)

Table 103 : Genetic effects for weight of 1000 seeds (g)

Cross combinations	Parameters m	d	H	I	j	l
$P_1 \times P_2$	1.64** ± 0.64	0.25** ± 0.01	5.27** ± 1.51	1.64* ± 0.64	0.34 ± 0.36	-4.14** ± 0.90
$P_1 \times P_3$	3.13** ± 0.68	0.52** ± 0.01	1.68 ± 1.64	-0.12 ± 0.68	-1.76** ± 0.40	-2.24* ± 0.98
$P_2 \times P_3$	2.02** ± 0.64	0.27** ± 0.01	4.43** ± 1.54	0.74 ± 0.64	0.56 ± 0.39	-3.42** ± 0.92
$P_2 \times P_1$	2.10* ± 0.83	-0.25** ± 0.01	5.33* ± 2.07	1.18 ± 0.83	1.60** ± 0.55	-4.42** ± 1.26
$P_3 \times P_1$	2.95** ± 0.64	-0.52** ± 0.01	1.45 ± 1.54	0.06 ± 0.63	1.82* ± 0.39	-1.22 ± 0.92
$P_3 \times P_2$	4.22** ± 0.66	-0.27** ± 0.01	-1.39 ± 1.51	-1.46* ± 0.66	1.56** ± 0.33	0.26 ± 0.87

(Source: Thirugnanakumar, 1991)

Table 104 : Genetic effects for volume of 1000 seeds (ml)

Cross combinations	Parameters m	d	H	I	j	l
$P_1 \times P_2$	7.14 ± 6.26	-0.52 ± 0.64	11.93 ± 15.51	0.16 ± 6.22	-3.92 ± 4.23	-12.90 ± 9.63
$P_1 \times P_3$	14.31* ± 5.98	-0.58 ± 0.55	-13.23 ± 13.16	-6.94 ± 5.95	-0.29 ± 2.59	7.03 ± 7.59
$P_2 \times P_3$	8.32** ± 0.59	0.02 ± 0.59	1.57 ± 1.18	-	-	-
$P_2 \times P_1$	3.64 ± 6.45	0.52 ± 0.64	15.62 ± 14.44	3.66 ± 6.42	1.70 ± 3.07	-12.16 ± 8.40
$P_3 \times P_1$	7.13 ± 6.03	0.58 ± 0.55	7.66 ± 13.96	0.24 ± 6.01	-4.97 ± 3.25	-7.27 ± 8.37
$P_3 \times P_2$	7.86** ± 0.59	-0.35 ± 0.55	0.34 ± 1.19	-	-	-

(Source: Thirugnanakumar, 1991)

Table 105 : Genetic effects for density of seeds (g/ml)

Cross combinations	Parameters m	d	H	I	j	l
P_1 ´ P_2	1.94* ± 0.81	0.04 ± 0.04	8.37** ± 2.23	2.48** ± 0.81	0.35 ± 0.68	-5.79** ± 1.44
P_1 P_3	0.53 ± 0.63	0.09** ± 0.02	1.76 ± 1.51	-0.05 ± 0.63	-0.36 ± 0.37	-1.82* ± 0.91
P_2 ´ P_3	0.48** ± 0.03	0.07* ± 0.03	0.22** ± 0.07	-	-	-
P_2 ´ P_1	0.53 ± 0.62	-0.04 ± 0.04	1.42 ± 1.46	0.01 ± 0.61	-0.14 ± 0.35	-1.35 ± 0.87
P_3 ´ P_1	-0.30 ± 1.11	-0.09** ± 0.02	4.86 ± 2.61	0.78 ± 1.11	-1.44* ± 0.62	-3.98* ± 1.55
P_3 ´ P_2	-1.65** ± 0.58	-0.05 ± 0.03	7.80** ± 1.57	2.10** ± 0.58	0.73 ± 0.47	-5.59** ± 1.01

* Significant at P = 5%; ** Significant at P = 1%

(Source: Thirugnanakumar, 1991)

Table 106 : Genetic effects for oil content (%)

Cross combinations	Parameters m	d	H	I	j	l
$P_1 \times P_2$	45.16** ± 3.08	1.49** ± 0.13	-5.62 ± 7.59	-6.80* ± 3.07	0.18 ± 2.00	1.20 ± 4.62
$P_1 \times P_3$	47.40** ± 3.18	1.01** ± 0.11	-16.27* ± 7.82	-8.56** ± 3.18	-4.71* ± 2.04	10.33* ± 4.75
$P_2 \times P_3$	47.31** ± 2.82	-0.48** ± 0.15	-13.58* ± 6.75	-9.96** ± 2.81	7.47** ± 1.69	9.37 ± 4.04
$P_2 \times P_1$	46.92** ± 3.29	-1.49** ± 0.13	-15.67 ± 8.51	-8.56** ± 3.29	2.94 ± 2.43	10.38 ± 5.34
$P_3 \times P_1$	38.74** ± 2.68	-1.01** ± 0.11	15.69* ± 6.27	0.10 ± 2.68	2.73 ± 1.46	-15.09** ± 3.79
$P_3 \times P_2$	35.35** ± 3.32	0.48** ± 0.15	21.92** ± 7.76	2.00 ± 3.31	-3.71* ± 1.82	-12.95** ± 4.58

(Source: Thirugnanakumar, 1991)

Table 107 : Genetic effects for seed yield (g)

Cross combinations	Parameters m	d	H	I	j	l
$P_1 \times P_2$	-11.54** ± 2.70	-0.33 ± 0.22	47.83** ± 7.08	13.88** ± 2.69	10.21** ± 2.08	-34.23** ± 4.49
$P_1 \times P_3$	1.11 ± 1.99	0.25 ± 0.19	11.55* ± 4.74	1.16 ± 1.98	2.30 ± 1.20	-11.18** ± 2.85
$P_2 \times P_3$	5.86** ± 1.60	0.08 ± 0.24	-4.62 ± 3.94	-3.26* ± 1.58	0.83 ± 1.10	0.71 ± 2.44
$P_2 \times P_1$	4.92** ± 1.62	0.33 ± 0.22	-1.38 ± 4.06	-2.58 ± 1.61	-0.71 ± 1.15	-1.43 ± 2.54
$P_3 \times P_1$	5.51** ± 1.57	0.25 ± 0.19	-3.61 ± 3.83	-3.24* ± 1.56	-1.10 ± 1.04	0.06 ± 2.35
$P_3 \times P_2$	0.53 ± 1.41	-0.08 ± 0.24	8.21* ± 3.62	2.06 ± 1.39	1.61 ± 1.09	-7.19** ± 2.31

(Source: Thirugnanakumar, 1991)

Table 108 : Genetic effects for total dry matter production (g)

Cross combinations	Parameters m	d	H	I	j	l
$P_1 \times P_2$	-33.12** ± 9.82	-2.55** ± 0.62	173.36** ± 23.67	49.68** ± 9.81	32.53** ± 5.98	-128.45** ± 14.26
$P_1 \times P_3$	6.93 ± 8.97	-1.33* ± 0.59	67.06** ± 21.23	8.42 ± 8.95	10.11 ± 5.16	-64.77** ± 12.61
$P_2 \times P_3$	45.31** ± 7.61	1.22* ± 0.58	-54.57** ± 17.82	-27.42** ± 7.59	4.70 ± 4.26	22.78* ± 10.52
$P_2 \times P_1$	37.97** ± 7.98	2.55** ± 0.62	-30.60 ± 19.56	-21.40** ± 7.96	-6.57 ± 5.15	6.93 ± 11.93
$P_3 \times P_1$	41.69** ± 8.27	1.33* ± 0.53	-41.65* ± 19.24	-26.34** ± 8.25	-6.07 ± 4.48	12.23 ± 11.29
$P_3 \times P_2$	10.87 ± 7.48	-1.22* ± 0.58	29.31 ± 17.93	7.02 ± 7.45	12.09** ± 4.53	-29.10** ± 10.79

(Source: Thirugnanakumar, 1991)

Table 109 : Genetic effects for harvest index (%)

Cross combinations	Parameters m	d	H	I	j	l
$P_1 \times P_2$	10.12 ± 6.86	0.68 ± 0.91	31.16 ± 16.69	4.34 ± 6.80	6.63 ± 4.50	-24.49* ± 10.32
$P_1 \times P_3$	19.79** ± 6.70	0.05 ± 0.85	-0.43 ± 16.45	-4.72 ± 6.69	-0.23 ± 4.47	-3.75 ± 10.30
$P_2 \times P_3$	15.25** ± 0.83	-0.54 ± 0.82	1.35 ± 1.59	-	-	-
$P_2 \times P_1$	15.39** ± 0.83	-0.60 ± 0.83	1.78 ± 1.52	-	-	-
$P_3 \times P_1$	15.61** ± 0.79	0.01 ± 0.78	1.77 ± 1.54	-	-	-
$P_3 \times P_2$	10.32 ± 6.87	0.62 ± 0.90	21.79 ± 16.72	4.08 ± 6.81	1.20 ± 4.50	-17.86 ± 10.42

* Significant at P = 5%; ** Significant at P = 1%

(Source: Thirugnanakumar, 1991)

Chapter 11

Genetic Analysis of Triple Test Cross Progenies

Among the different biometrical methods available, triple test cross analysis is one of the most efficient designs currently available for investigating the genetic structure of populations. It provides a test of epistasis and in its absence it gives independent and equally precise estimates of additive and dominance genetic components (Mather and Jinks, 1982).

11.1. Statistical analysis

Triple test cross

Jinks and Perkins (1970) method was applied to detect epistasis and estimate additive and dominance components of genetic variance.

Test for epistasis

For the test of epistasis, twenty values of $L_{1i} + L_{2i} - 2L_{3i}$ for i = 1 to 20 were obtained for each of the three replicates. After summing over the replicates, the sum of twenty squred deviations of $L_{1i} + L_{2i} - L_{3i}$ fromn zero for 20 degrees of freedom, that is sum of squares due to epistasis was obtained.

Sum of squares due to replicate error, that is the error for testing the significances of epistasis, was obtained for 40 degrees of freedom.

A further estimate of the error variance was obtained by the mean variance withitn the L_1, L_2 and L_3 types of families for 720 degrees of freedom.

This analysis was, however, pursued further the epistasis sum of squares for 20 degrees of freedom could be partitioned into an item for one degree of freedom, testing for 'i' type epistasis (additive additive interactions) and an item for 19 degrees of freedom testing for (j) and (l) types of epistasis (additive × dominance and dominance × dominance interactions respectively).

Test and estimation of additive and dominance components

Additive and dominance components were estimated assuming no epistasis. The twenty sums of means of the families yielded a variance of sums for19 degrees of freedom. Similarly, the variance of differences was obtained further for 19 degrees of freedom.

Item	Df	EMS
Sums ($L_{1i} + L_{2i} + L_3$)	n-1	$\sigma^2 + 3r\sigma^2 m$
Differences ($L_{1i} - L_{2i}$)	n-1	$\sigma^2 + 2r\sigma^2 ml$
Error	n(r-1)	σ^2

(Source: Jinks and Perkins, 1970)

For the F_2 and summed items from the two back cross populations the expectations for s^2m and s^2ml in the absence of epistasis are

$$\text{No linkage } \sigma^2_m = \frac{1}{8}\sum d^2$$

$$\sigma^2_{ml} = \frac{1}{8}\sum h^2$$

$$\text{Linkage } s^2_m = \frac{1}{8}\sum d^2 + \frac{1}{4}\sum(1\text{-}2P_jK)d_id_k$$

$$\sigma^2_{ml} = \frac{1}{8}\Sigma h^2 + \frac{1}{4}\Sigma(1-2P_hK)h_ih_k$$

In the presence of linkage, D equals $\Sigma d^2 \pm \Sigma(1\text{-}2P_jK)d_id_k$ where + is for coupling linkage and the – is for repulsion linkage P_jK is the recombination between j^{th} and k^{th} loci and H equal $\Sigma h^2 + 2\Sigma(1\text{-}$

$2P_jK) h_i h_k$. In the absence of linkage, D the additive components equals Σd^2 and H the dominance component equals Σh^2.

11.2. Example

Saravanan (2001) made triple test cross analysis with the three triple test cross combinations *viz.*, (1) Arka Anamika × Parbani Kranti, (2) Arka Anamika × MDU 1 and (3) Parbani Kranti × MDU 1 (Tables 110 and 111).

11.2.1. Triple test cross analysis

Componentw of variance

The results of the analysis of variance to detect additive, dominance componets and epistatsis for all nine characters in three triple test crosses are presented in Tables 112-121.

Days to first flower

The mean sum of squares for additive components were significant in crosses 1 and 3. While, dominance components were significant in crosses 2 and 3. Regarding epistasis, total epistasis and 'j' and 'l' types were significant for this trait in cross 2. The 'i' type epistasis was not significant in all the crosses for this character (Table 112).

Plant height

The additive and dominance components were highly significant in all the three crosses for the plant height. The total epistasis was highly significant in all the crosses. Further partitioning of the epistasis however, revealed that the 'i' type epistasis was significant in cross 1. But all the crosses had significant 'j' and 'l' type of epistasis for this character (Table 113).

Table 110 : Analysis of variance to test for epistasis in three triple test crosses *viz.*, 1 (Arka Anamika ´ Parbani Kranti), 2. (Arka Anamika ´ MDU 1) and 3. (Parbhani Kranti ´ MDU 1)

Item	Crosses	df	Mean sum of squares								
			Days to first flower	Plant height	Inter node length	No. of nodes per plant	No. of fruits per plant	Fruit length	Fruit girth	Fruit weight	Fruit yield per plant
Epistasis	1	20	1.67	237.13**	1.52*	18.72*	10.46*	2.62*	0.80**	35.38	13160.71**
$(L_{1i} + L_{2i} - L_{3j})$	2	20	2.11*	860.45**	1.03*	55.04**	57.18**	15.23**	0.62**	61.45*	10670.31**
	3	20	1.20	1017.10**	4.71**	62.02**	52.10**	4.88**	0.42*	44.69**	7447.76**
Over all	1	1	2.18	1439.20**	1.60	5.67	8.40	0.79	0.02	37.22	13327.82**
epistasis	2	1	0.48	68.64	1.54	15.71	28.56*	15.24**	0.79	63.93**	11037.48**
(i type)	3	1	1.52	64.52	1.51	1.32	1.20	2.55	0.03	46.98**	7780.65**
Epistasis	1	19	1.75	173.86**	1.51*	19.35*	10.60	2.71	0.84	0.59	9985.60**
(j and i type)	2	19	2.19*	902.13**	1.00*	57.11**	58.69**	15.23**	0.61	14.19	3694.08
	3	19	1.19	1067.23**	4.87**	65.22**	54.78**	5.01**	0.43	1.17	1122.89
Replicate	1	40	0.31	67.99	0.53	5.85	2.07	1.15	0.06	11.03	1101.27
error	2	40	0.47	60.90	0.36	7.49	4.46	2.62	0.19	12.50	417.49
	3	40	0.21	4.10	0.02	0.15	0.21	0.09	0.01	1.93	416.59
Replicates	1	2	2.53	19315.11**	41.89**	198.50**	414.15**	26.37*	6.38**	122.35	20380.60**
	2	2	3.20	6468.30**	3.93*	68.91	6.16	10.79	1.07	51.43	74487.92**
	3	2	2.12	271.41**	0.54**	52.62**	65.36**	0.71**	0.57**	35.58**	8734.77**
Epistasis ×	1	38	0.32	63.81	0.53	5.99	2.17	1.13	0.06	11.27	1139.57
Replicates	2	38	0.49	62.62	0.31	7.17	3.78	2.74	0.21	10.28	423.63
	3	38	0.22	4.31	0.02	0.16	0.19	0.09	0.01	1.89	441.62
	1	720	1.95	95.59	1.47	16.90	8.83	2.42	0.33	39.17	1811.15
	2	720	1.62	129.93	0.73	12.66	8.80	1.66	0.28	37.80	2392.69
	3	720	1.64	175.42	1.01	10.96	7.04	1.45	0.32	24.71	2813.03

*Significant at 5%; ** Significant at 1%

(Source: Saravanan, 2001)

Table 111 : Analysis of variance to detect additive and dominance components in three crosses *viz.*, 1 (Arka Anamika × Parbani Kranti), 2. (Arka Anamika × MDU 1) and 3. (Parbhani Kranti × MDU 1)

Item	Crosses	df	Mean sum of squares								
			Days to first flower	Plant height	Inter node length	No. of nodes per plant	No. of fruits per plant	Fruit length	Fruit girth	Fruit weight	Fruit yield per plant
Analysis of sums($L_{1i}+L_{2i}-L_{3j}$)											
Epistasis	1	19	3.45**	501.26**	3.36**	52.00**	43.71**	2.96**	0.48	61.16**	11096.31**
(additive	2	19	0.77	498.67**	5.19**	71.39**	6.38**	6.38**	0.69	144.10**	12374.51**
component)	3	19	3.34**	2187.43**	8.22**	149.35**	5.53**	5.53**	0.50	195.98**	22641.06**
	1	38	0.27	43.27**	0.85	9.44**	1.15	1.15	0.08	26.26**	2035.73**
	2	38	0.48	52.39**	0.43	21.96**	1.29	1.29	0.17	26.35**	828.70**
	3	38	0.22	13.44**	0.02	0.31	0.04	0.04	0.01	3.32**	351.69**
Families	1	720	1.95	95.59	1.47	16.91	2.42	2.42	0.33	39.17	1811.10
	2	720	1.62	129.93	0.73	12.66	1.66	1.66	0.28	37.80	2392.69
	3	720	1.64	175.43	1.01	10.96	1.45	1.45	0.32	24.72	2813.04
Analysis of difference ($L_{1i}-L_{2i}$)											
Differences	1	19	0.75	291.38**	2.47**	24.53**	3.16**	3.16**	0.43	37.86**	7954.07**
dominance	2	19	1.96*	808.94**	2.02*	68.40**	7.28**	7.28**	0.59	25.49**	23107.45**
component	3	19	1.97*	1314.40**	3.61**	29.60**	4.45**	4.45**	0.39	33.93**	13895.48**
Differences	1	38	0.26	39.87**	0.47	8.86**	1.90**	1.90**	0.08	14.66**	1487.61**
× replicates	2	38	0.39	54.83**	0.30	5.74**	1.12	1.12	0.19	9.99**	892.18**
	3	38	0.26	11.82**	0.01	0.17	0.02	0.02	0.01	1.46**	267.45**
Within	1	480	2.04	91.87	1.60	16.18	2.29	2.29	0.34	37.72	1877.14
families	2	480	1.78	111.46	0.82	13.59	1.48	1.48	0.27	19.60	2500.11
	3	480	1.64	189.20	1.18	10.32	1.53	1.53	0.36	18.51	2644.01

*Significant at 5%; ** Significant at 1%

(Source: Saravanan, 2001)

Table 112 : Components of variance for days to first flower in bhendi crosses

Components	Cross - 1	Cross - 2	Cross - 3
D	3.45**	0.77	3.34**
H	0.75	1.96*	1.97*
Total epistasis	1.67	2.11*	1.20
i type epistasis	2.18	0.48	1.52
j and l type epistasis	1.75	2.19**	1.19

*Significant at 5 %; ** Significant at 1 % (Source: Saravanan, 2001)

Table 113 : Components of variance for plant height in bhendi crosses

Components	Cross - 1	Cross - 2	Cross - 3
D	501.26**	498.67**	2187.43**
H	291.38**	808.94**	1314.40**
Total epistasis	237.13**	860.45**	1017.10**
i type epistasis	1439.20**	68.64	64.52
j and l type epistasis	173.86**	902.13**	1067.23**

*Significant at 5 %; ** Significant at 1 % (Source: Saravanan, 2001)

Inter node length

The variances due to additive and dominance components were highly significant for inter node length in all the three crosses. All the three crosses had highly significant total epistasis for inter node length. The 'j' and 'l' type epistasis were highly significant in all the crosses for this character. However 'i' type of interaction was not significant in all the three crosses (Table 114).

Table 114 : Components of variance for inter node length in bhendi crosses

Components	Cross - 1	Cross - 2	Cross - 3
D	3.36**	5.19**	8.22**
H	2.47**	2.02*	3.61**
Total epistasis	1.52*	1.03*	4.71**
i type epistasis	1.60	1.54	1.51
j and l type epistasis	1.51*	1.00*	4.87**

*Significant at 5 %; ** Significant at 1 % (Source: Saravanan, 2001)

Number of nodes per plant

The variances due to additive and dominance components were highly significant in all the crosses. The total 'j' and 'l' types of epistasis were highly significant in all the crosses. But the 'i' type epistasis was non-significant in all the crosses for number of nodes per plant (Table 115).

Table 115 : Components of variance for number of nodes per plant in bhendi crosses

Components	Cross - 1	Cross - 2	Cross - 3
D	52.00**	71.39**	149.35**
H	24.53**	68.40**	29.60**
Total epistasis	18.72*	55.04**	62.02**
i type epistasis	6.67	15.71	1.32
j and l type epistasis	19.35*	57.11**	65.22**

*Significant at 5 %; ** Significant at 1 % (Source: Saravanan, 2001)

Number of fruits per plant

The variances of mean squares for additive and dominance components were significant in all the three crosses. The total epistasis was highly significant in all the crosses. Further partitioning of the epistasis however, revealed that the 'i' type epistasis was significant in cross 2 and non-significant in crosses 1 and 3. The highly significant 'j' and 'l' type of epistasis were observed in crosses 2 and 3 for number of fruits per plant (Table 116).

Table 116 : Components of variance for number of fruits per plant in bhendi crosses

Components	Cross - 1	Cross - 2	Cross - 3
D	43.71**	34.17**	124.75**
H	10.85**	60.79**	30.33**
Total epistasis	10.49*	57.18**	52.10**
i type epistasis	8.40	28.56*	1.20
j and l type epistasis	10.60	58.69**	54.78**

*Significant at 5 %; ** Significant at 1 % (Source: Saravanan, 2001)

Fruit length

The variances due to additive and dominance components were highly significant in all the three crosses. The total epistasis was highly significant for fruit length in all the crosses. The cross 2 only recorded highly significant 'i' type epistasis. However, the cross 2 and 3 recorded significant 'j' and 'l' type epistasis for this trait (Table 117).

Table 117 : Components of variance for fruit length in bhendi crosses

Components	Cross - 1	Cross - 2	Cross - 3
D	2.96**	6.38**	5.53**
H	3.16**	7.28**	4.45**
Total epistasis	2.62*	15.23**	4.88**
i type epistasis	0.79	15.24**	2.55
j and l type epistasis	2.71	15.23**	5.01**

*Significant at 5 %; ** Significant at 1 % (Source: Saravanan, 2001)

Fruit girth

The variances due to additive and dominance components were non-significant in all the three crosses. All interaction effects also non-significant in all the crosses for fruit girth (Table 118).

Table 118 : Components of variance for inter node length in bhendi crosses

Components	Cross - 1	Cross - 2	Cross - 3
D	0.48	0.69	0.50
H	0.43	0.59	0.39
Total epistasis	0.80	0.62	0.42
i type epistasis	0.02	0.79	0.03
j and l type epistasis	0.84	0.61	0.43

(Source: Saravanan, 2001)

Fruit weight

The mean squares for additive as well as dominance components were highly significant in all the crosses. The crosses 2 and 3 recorded highly significant total epistasis and 'i' type epistasis for fruit weight.

However, the 'j' and 'l' type of epistasis was non-significant in all the crosses for this trait (Table 119).

Table 119 : Components of variance for fruit weight in bhendi crosses

Components	Cross - 1	Cross - 2	Cross - 3
D	61.16**	144.10**	195.98**
H	37.86**	25.49**	33.93**
Total epistasis	35.38	61.45*	44.69**
i type epistasis	37.22	63.93*	46.98**
j and l type epistasis	0.59	14.19	1.17

*Significant at 5 %; ** Significant at 1 % (Source: Saravanan, 2001)

Fruit yield per plant

The variances for additive and dominance components were highly significant in all the crosses. The total epistasis and 'i' type epistasis were highly significant for this trait in all the crosses. However, highly significant 'j' and 'l' type of epistasis was recorded in cross 1 only (Table 120).

Table 120 : Components of variance for fruit yield per plant in bhendi crosses

Components	Cross – 1	Cross - 2	Cross - 3
D	11096.31**	12374.51**	22641.06**
H	7954.07**	23107.45**	13895.48**
Total epistasis	13160.71**	10670.31**	7447.76**
i type epistasis	13327.82**	11037.48**	7780.65**
j and l type epistasis	9985.60**	3694.08	1122.89

*Significant at 5 %; ** Significant at 1 % (Source: Saravanan, 2001)

11.2.2. Genetic components

The estimates of additive (D) and dominance (H) components for nine characters are presented in Table 121.

Days to first flower

The additive 'D' component and the dominance 'H' component were highly significant in crosses 1 and 3. 'D' component was higher than 'H' component. The values for 'D' were 1.33 and 1.51 and for 'H' were 0.65 and 0.44 respectively for the two crosses. The degree of dominance $(H/D)^{1/2}$ was less than unity in all the crosses.

Table 121 : Genetic components for nine characters in bhendi crosses

Components	Components	Cross - 1	Cross - 2	Cross - 3
Days to first flower	D	1.33*	0.76*	741.51**
	H	0.65*	0.24	0.44**
	$(H/D)^{1/2}$	0.69	0.56	0.54
Plant height	D	360.59**	327.77**	1788.44**
	H	266.01**	929.97**	1500.27**
	$(H/D)^{1/2}$	0.86	1.68	0.92
Inter node length	D	1.68**	3.96**	6.41**
	H	1.16**	1.60	3.24*
	$(H/D)^{1/2}$	0.83	0.63	0.71
Number of nodes per plant	D	31.19**	52.20**	123.01**
	H	11.13**	73.08**	25.71**
	$(H/D)^{1/2}$	0.19	1.18	0.46
Number of fruits per plant	D	31.00**	22.55**	104.55**
	H	3.59**	68.01**	31.77**
	$(H/D)^{1/2}$	0.34	1.74	0.55
Fruit length	D	0.48	4.19**	3.63**
	H	1.16*	7.73**	3.89**
	$(H/D)^{1/2}$	1.55	1.36	1.03
Fruit girth	D	0.13*	0.36**	0.16*
	H	0.12	0.43**	0.04
	$(H/D)^{1/2}$	0.52	1.09	0.5
Fruit weight	D	19.55*	94.49**	152.23**
	H	0.19	7.85**	20.56**
	$(H/D)^{1/2}$	0.09	0.29	0.37
Fruit yield per plant	D	8253.52**	8872.73**	17624.90**
	H	8102.57**	27476.45**	15001.96**
	$(H/D)^{1/2}$	0.99	1.76	0.92

*Significant at 5 %; ** Significant at 1 % (Source: Saravanan, 2001)

Plant height

The additive 'D' component and the dominance 'H' component were highly significant in all the three crosses. The estimate for additive 'D' component ranged from 327.77 (cross 2) to 1788.44 (cross 3) and the dominance 'H' component varied from 266.01 (cross 1) to 1500.27 (cross 3). The degree of dominance $(H/D)^{1/2}$ was cross 2. The degree of dominance $(H/D)^{1/2}$ was more than unity in cross 2 (1.68) and it was less than unity in the remaining two crosses.

Inter node length

The 'D' and 'H' components were highly significant in crosses j and 3. The additive component ranged from 1.68 (cross 1) to 6.41 (cross 3) and dominance component had a variation from 1.16 (cross 1) to 6.44 (cross 3). The degree of dominance represented by $(H/D)^{1/2}$ was less than unity in all the crosses.

Number of nodes per plant

Through 'D' and 'H' components were highly significant and 'D' was higher than 'H' in all the crosses. The 'D' component ranged from 31.19 in cross 1 to 123.01 in cross 2 and H component from 11.13 in cross 1 to 73.08 in cross 2. The degree of dominance $(H/D)^{1/2}$ was more than unity in cross 1 (1.18) and it was less than unity in other two crosses.

Number of fruits per plant

The additive 'D' component and the dominance 'H' component were highly significant in all the crosses. Additive component was more than dominance component. The estimate for additive 'D' component ranged from 22.55 in cross 2 to 104.55 in cross 3 and the dominance 'H' component varied from 3.59 in cross 1 to 68.01 in cross 2. The degree of dominance $(H/D)^{1/2}$ was more than unity in cross 2 and it was less than unity in the remaining two crosses.

Fruit length

The estimate of 'D' was significant in cross 2 (4.19) and cross 3 (3.63). The 'H' component was significant in all the crosses. The degree of dominance $(H/D)^{1/2}$ was more than unity in all the three crosses.

Fruit girth

The additive 'D' component was significant in all the crosses. The dominance 'H' was significant in cross 2 (0.43). The degree of dominance $(H/D)^{1/2}$ was more than unity in cross 2 (1.09) and in other two crosses *viz.*, cross 1 (0.52) and cross 3 (0.5) it was less than unity.

Fruit weight

The 'D' component was highly significant in all the crosses. While 'H' component were significant in crosses 2 and 3 (7.85 and 20.56 respectively). The degree of dominance represented by $(H/D)^{1/2}$ was less than unity in all the crosses.

Fruit yield per plant

The 'D' and 'H' components were highly significant in all the crosses. The 'D' component ranged from 8253.52 in cross 1 to 17624.9 in cross 3 and 'H' component from 8102.57 in cross 1 to 27476.45 in cross 2. The degree of dominance $(H/D)^{1/2}$ was more than unity in only cross 2 (1.76) and in other two crosses 1 and 3 was less than unity.

Chapter 12

Genetic Analysis of Biparental Progenies

Generation of new genetic variability through recombination breeding followed by appropriate selection procedure is one of the key factor for crop improvement. Gain in selection is restricted by correlated response of agronomically desirable as well as undesirable characters. There is a host of examples in various crop species where genes for desirable agronomic traits have been found to be closely linked with those governing undesirable characters (Murthy, 1971; Clegg *et al.*, 1972). Such close linkages delay the realization of the full recombination potential in many crops. The biparental mating or the disruptive mating scheme, finds its importance in the improvement of autgamous speice,s because it increases the probabilities of obtaining useful recombinants as compared to selection without intermating (Hanson, 1959; Jensen, 1970).

In contrast to classical pedigree breeding, random mating population could produce large amounts of variability. This variability can be maintained in the population by the use of recurrent intermating, while the population is improved by selection. According to the standard breeding methodology in self-pollinated crops selection for quantitative characters are generally taken up in early segregating generations. For characters like yield, selection is continued till the material becomes homozygous. Such characters are controlled by large number of genes and a very large population has to be raised to make selection effective. Since most plant breeders could never grow the theoretically required size of population, the recombination is highly restricted. It is likely that many of the valuable genes may be practically lost in the process of advancing the material by continued selfing the generations. Andrus (1963) suggested crossing of selected sibs in early generation for reassembling the genes capable of functioning in a balanced polygenic system.

Among the methods of increasing the frequency of desirable recombinants the system of biparental mating is reported to alter the phase of linkage through forced recombinations. A greater amount of concealed genetic variation, particularly of additive type, is said to be released. Such a mating system therefore could be expected to improve response to selection. Intercrossing in the segregating generatins more variations and causes alterations in the genetic correlations between several characters in self-pollinated crops (Miller and Rawlings, 1967; Matzinger and Wernsman, 1968; Gill *et al.*, 1973; Redden and Jensen, 1974; Verma *et al.*, 1979; Yadav and Murthy, 1979).

Moll and Robinson (1967) and Miller and Rawlings (1967) considered that the estimates of additive genetic component from advanced generations would be more reliable than those of from corresponding F_2 generation. They suggested that estimate from the F_2 might be biased in the presence of repulsion phase linkages which over estimate non-additive variance and therefore conceal additive variance. If this proportion holds true, superior recombinants should be expected in the progenies of biparental matings compared with selfed progenies.

Redden and Jenson (1974) after investigating the Bips in barley and wheat, explained that intermating the plants of segregating population as a good tool for breeding programmes for naturally inbreeding crops provided that the additive component of the genetic variance found to be present. Joshi (1979) has suggested that intermating of selected plants in F_2 was the best technique so as to accumulate the favourable genetic effects and to break linkages thereby to enable the plant breeders to exercise selection procedure.

Hanson *et al.* (1967) and Gill *et al.* (1973) obtained encouraging results due to biparental mating in the segregating generations of self-pollinated crops. This was later confirmed by Yadav and Murthy (1979) and Vermal *et al.* (1979) in wheat.

12.1. Statistical analysis

12.1.1. North Carolina design–I

The data for each character were analysed on the basis of analysis of variance techniques. Each character was analysed separately overall the sets. The appropriates variance analysis used for each character is outlined as follows.

ANOVA for North Carolina Design – I

Source of variance	Degrees of freedom	Mean squares	Expected mean squares
Sets	s-1		
Replication in sets	s (r-1)		
Males in sets (M/S)	s (m – 1)	M_1	$\sigma^2_e + n\sigma^2_P + nr\sigma^2_f + nf\sigma^2_m$
Females in males in sets (F/M/S)	Sm (f-1)	M_2	$\sigma^2_e + n\sigma^2_P + nr\sigma^2_f$
Replication ´ Females	s (mf-1) (r-)	M_3	$\sigma^2_e + n\sigma^2_P$
Error	Smfr (n-1)	M_4	σ^2_e

(Source: Comstock and Robinson, 1952)

where,

s = mean of sets

r = number of replication per set

m = number of males pet set

f = number of females per male

s^2_e = Error variance

s^2_P = Variance of plot effects

s^2_m = Variance of progeny of different males mated to different females.

s^2_f = Variance of progeny families of different females

n = number of plants on which observation taken per fullsib family.

By using the appropriate mean square obtained from the analysis of variance, an estimate of the variance among males (s^2_m) and variance among females (s^2_f) mated to the same male were obtained as follows.

$\sigma^2_m = (M_1 - M_2)/rfn$

$\sigma^2_f = (M_2 - M_3)/rf$

12.2. Example

Deepalakshmi (2002) studied the genetics of seed yield and its component characters in biparental progenies through North Carolina Design I in sesame. The experimental material was constituted by seeds to two cross combinations and their F_1 generation. The particulars of the cross combination and the generation studied are furnished below.

Cross combinations

Cross 1	CO 1 × AUS 107
Cross 2	Paiyr 1 × AUS 107

Generations studied

1. F_2 generation
2. F_3 generation
3. Bips generation

12.2.1. Mean performance

Mean is a basic and an important criterion in selecting superior segregants. According to Finkner *et al.* (1973) progenies with the highest mean was relatively effective in selecting the superior segregants.

Joshi (1979) opined that intermating of F_2 population has been found to boost up population means in biparental progenies (Bips). Yunus and Paroda (1983) reported the superiority of Bips over F_3 in terms of mean performance in wheat. Such superior performance of intermated progenies is of immense value as population means go on decreasing progressively from F_2 generations onwards the homozygosity increases from F_3 onwards. Pant *et al.* (1992) suggested that the superior means in the Bips may be primarily due to the generation of more genetic variability in the Bips than in the F_3. The mean of F_3 and biparental progenies should be the same in the absence of linkage (Mather and Jinks, 1971).

Bips of both crosses exhibited higher mean values at high frequency for almost all traits. The mean performance of Bips was in general, slightly better than F_3 progenies for all the characters in both the crosses. It was also interesting to find that mean performance improved considerably for seed yield per plant in both the crosses through intermating compared to the F_3 progenies (Table 122).

Table 122 : Mean performance for seed yield per plant in F_2, F_3 and Bips populations

Population	CO 1 × AUS 107		Paiyur1 × AUS 107	
	Range (g)	Mean (g)	Range (g)	Mean (g)
F_2	2.7–18.4	6.83	1.8–16.7	5.77
F_3	2.4–16.9	6.24	1.6–16.1	5.86
Bips	3.1–23.9	7.09	1.6–17.5	6.59

(Source: Deepalakshmi, 2002)

Increasing in mean values in Bips for many traits was already reported (Gurdev Sing *et al.*, 1986; Nanda *et al.*, 1990) in wheat; (Muthukumaran, 1993) in sesame and (Vijayakumar, 1998) in cotton.

Increase in mean values for many of the characters in the biparental progenies as compared to F_3 progenies can be explained, if we consider that the Bips are the immediate progenies of a cross between selected plants in the F_2 generation, whereas the F_3 progenies were derived by selfing F_2 plants. The heterozygosity in Bips may confer slight advantage as compared to F_3. When compared with F_2 mean values, an increase was noticed in Bips (Gardner *et al.*, 1953).

Matzinger *et al.* (1960) and Matzinger and Wernsman (1968) reported higher mean among full-sib progenies over F_3 for majority of the characters in tobacco. Bains (1971) observed that the family means of Bips in cotton in terms of yield per plant were not only comparable but also surpassed the corresponding F_2 mean and in one case the Bips mean actually approached the F_1 mean. He had attributed that it might be due to the accumulation of additive as well as complementary epistatic effects.

Hence, the Bips developed in both the crosses can be utilized further and they can serve as a base population for improvement of sesame yield as they recorded higher mean values for seed yield per plant.

Oil content is the ultimate factors of high productivity of oilseeds crop like sesame. Oil population can be improved through increased seed yield or increased oil content of the seed. In sesame oil content ranges from 40-60 per cent depend upon the seed colour and texture. In the present study the mean performance of oil content was 50.58 and 50.46 per cent in Bips population of cross 1 and cross 2 respectively. There was a wide range of variation in oil content (47.90-52.99 per cent) in Bips population of cross 2.

Parameswari *et al.* (1998) reported that the intermating population exhibited slight improvement (0.4 per cent) in oil content of the cross CO 1 × TMV 6. Hence intermitted progenies may be considered sutiable for the selection of superior genotypes for high oil content when compared to F_3 population.

Table 123 : Mean performance for oil content in F_2, F_3 and Bips populations

Population	CO 1 × AUS 107		Paiyur1 × AUS 107	
	Range (g)	Mean (g)	Range (g)	Mean (g)
F_2	48.81–52.60	50.46	38.23–49.79	48.02
F_3	47.25–52.25	49.98	42.37–51.88	48.43
Bips	47.90–52.99	50.58	48.64–54.26	50.46

(Source: Deepalakshmi, 2002)

12.2.2. Gene action

Estimates of additive variance was high in CO 1 × Aus 107, whereas in Paiyur 1 × AUS 107, dominance variance was higher for seed yield (Table 124).

Table 124 : Estimation of variance components of Bips for seed yield per plant

Variance components	CO 1 × AUS 107	Paiyur 1 AUS 107
Additive	0.58	0.72
Dominance	0.34	1.29
Degree of dominance	0.60	1.89

(Source: Deepalakshmi, 2002)

Gardner *et al.* (1953) estimated the average degree of dominance for genes determining quantitative characters. The biparental progenies of cross 1 established substantial evidence for presence of over dominance (a > 1.0) for the characters like plant height, number of branches per plant, number of capsules per plant, number of seeds per capsule and 1000 seed weight, whereas seed yield per plant showed partial dominance (a < 1). The Bips population of cross 2 showed presence of over dominance for many of the characters studied.

The dominance variance was predominant even though additive variance was also present. For the exploitation of both gene action, intermatin of segregating population followed by reciprocal recurrent selection is suggested.

Moll and Stuber (1971) stated that reciprocal recurrent selection is designed to capitalize upon gene combinations that enhance hybrid performance, such as many loci showing complete dominance, over dominance and certain kinds of epistasis. Umakanth *et al.* (2000) reported in his studies, considerable amount of dominance variance along with additive variance was found to be governing most of the characters.

12.2.3. Variability

According to Allard (1960) variability indicates the extent of recombination for effective selection. Biparental intermating in early segregating generations in self-pollinated crops was recommended to generate and retain greater variability for several cycles of selection and elevate population mean (Hanson, 1959; Miller and Rawling, 1967). Sivasubramanian and Madhava Menon (1973) suggested high genetic variability as a pre-requisite for effective selection.

Bains (1971) revealed the evidences of inter-crossing in the segregating generations in generating more variations. Sharma *et al.* (1979) suggested that Bips develop rare recombinants, which remain restricted due to linkage, thereby it rapidly generate variability.

Intermating in F_2 generates more variability by breaking undesirable linkages and is only useful for the characters that exhibit repulsion phase linkages. The superior performance of Bips could be primarily attributed to the possible accumulation of favourable genes of low frequency. This makes feasible the emergence of those segregants which would have been rarely obtained in the simple F_2 and F_3 breeding populations. The intermating also generated more variability through offering an additional opportunity for genetic recombinations to occur while making random matings of F_2 plants (Gill *et al.*, 1973; Yunus and Paroda, 1983; Gurdevv Singh *et al.*, 1986).

Krishnadoss and Kadambavanasundaram (1987) also suggested that biparental crossing in sesame can improve yield by recurrent selection on the variability generated.

Variability in a population is measured by the estimates like phenotypic and genotypic coefficient of variations (Allard, 1960). While considering the variability in terms of genotypic coefficient of variation in the present investigation, Bips generated a higher variability when compared to F_3 population.

Plant height, number of capsules per plant and number of seeds per capsule in cross 1; number of capsules per plant, number of seeds per capsule and seed yield per plant in cross 2 showed the high variability in the Bips population. This substantiated the fact that the increase in genetic variability which is not available in F_3 generation was realized in Bips due to biparental mating in these traits.

Khadr and Frey (1965) observed increased genetic variances through intermating in oats. Similar results were obtained by Shanthi (1989) in rice and Parameswari *et al.* (1998) in sesame.

However, some of the traits such as number of branches per plant and seed yield per plant in cross 1 and plant height in cross 2 showed lesser variability in Bips when compared to F_3 population.

This was similar to the findings of Altman and Busch (1984) who observed that recombinants obtained through intermating of F_2 segregants did not enhance the genetic variances as many observed in spring wheat and they suggested that all intermatings could not develop enough variability. Hence they suggested the need for two or more cycles of intermating and selection among the progenies to create more variability.

Further, crosses with predominant coupling phase linkages would be expected to have reduced the genetic variance upon intermating. Genetic variance would be expected to increase for those crosses with a preponderance of repulsion phase linkages (Gardner, 1963; Singh and Murthy, 1973).

Nevertheless, considering the high variability created in Bips for number of capsules per plant and seed yield per plant in both the crosses in the present study, intermating of F_2 segregants is an effective tool for creating high variability and effective selection.

Therefore, the Bips developed in the present study will serve as good source for selection and improvement of many traits in sesame.

Thus, Bips not only increased the mean of the population but also enhanced some genetic variability in the present study, so this enhanced variability could be utilized for effective selection (Dwivedi and Singh, 1980). The superior performance of biparental populations in respect of genetic variation over F_2 base population and F_3 selected families was also expected from the release of concealed variability by breakage of undesirable linkages and by even some role of dominance variances of the biparental progenies (Poorn Chand and Raghunadha Rao, 2001).

Table 125 : Genotypic and phenotypic coefficient of variation for seed yield per plant in F_2, F_3 and Bips populations

Population	CO 1 × AUS 107		Paiyur1 × AUS 107	
	GCV (%)	PCV (%)	GCV (%)	PCV (%)
F_2	21.67	42.18	29.41	38.79
F_3	25.74	43.95	31.56	45.15
Bips	24.39	35.40	35.88	46.23

(Source: Deepalakshmi, 2002)

12.2.4. Heritability and genetic advance

Heritability estimates should be considered in terms of selection concept (Hanson, 1959). Murthy and Ziauddin Ahmad (1974) reported that the tight linkages may not be broken under selfing, while the mating system in Bips can break them. Therefore there was only linked change in heritability values from F_2 to F_3 but substantial increase in Bips.

Sharma *et al.* (1979) reported that the estimates of heritability obtained from Bips ought to be more realistic than those from F_2 generations which carries the load of repulsion phase linkages causing upward bias in the dominance variance.

Genetic advance being the product of heritability and selection differential, indicates the potentiality of selection intensity. Moreover, genetic advance considered along with heritability gives a reliable assessment of the resultant effect of selection in breeding populations. The estimates of high heritability do not always signify an increased genetic advance (Johnson *et al.*, 1955).

A comparison of heritability estimates between the Bips and F_3 population revealed that heritability estimates improved in intermated population in both the crosses indicating that Bips possessed large amount of heritable variance compared to F_3.

Andrus (1963) reported higher heritability estimates in case of Bips as compared to selfed progenies for some of the characters in wheat. High heritability in the case of Bips over that of F_2 and F_3 has also been reported by Gill *et al.* (1973) and Yunus and Paroda (1983) in wheat, Parameswari *et al.* (1998) in sesame and Pooran Chand and Raghunadha Rao (2001) in blackgram.

High heritability coupled with high genetic advance was observed for plant height and number of seeds per capsule in cross 1; number of branches per plant and seed yield per plant in cross 2. It is indicative of additive gene action and simple phenotypic selection may be practiced to improve the character. According to Johnson *et al.* (1995) existence of high heritability along with high genetic advance was due to additive gene action (Table 125).

12.2.5. Correlation studies

Quantitative characters are complexly inherited and coefficient of correlation values indicate the intensity and direction of character association in a crop. The interrelationship of component characters of yield provide the information about the likely consequence of selection for simultaneous improvement of desirable characters under selection. The release of variation as a result of recombination can also be examined by studying the interrelationships between different characters in the population.

Previous workers who studied biparental mating in crop plants had suggested biparental mating as a very important technique through which desirable character associations could be developed.

Nanda *et al.* (1990) revealed that the magnitude of correlation coefficients is expected to increase if linkage was in a predominantly repulsion phase. Intermating preserves genetic variability in the population and reduces the chances of genetic drift and unfavourable correlated response.

Pant *et al.* (1992) observed that the intermating frequently increased the genetic recombinations and changed the nature and magnitude of character associations in bread wheat and forage oats, respectively.

12.2.5.1. Correlation in F_2 population

In F_2 generations plant height, number of branches per plant and number of capsules per plant were found to be positively and significantly associated with seed yield per plant whereas, seed yield per plant was found to be negatively and significantly correlated with number of seeds per capsule and 1000 seed weight in cross 1. In cross 2, seed yield per plant was significantly positively correlated with number of branches per plant and number of capsules per plant whereas it was negatively correlated with number of seeds per capsule.

12.2.5.2. Correlation in F_3 and Bips populations

In the present study a shift from negative significant correlation to significant positive correlation was observed from F_3 to Bips. For exmple a shift from significant negative to significant positive correlation was observed for plant height and number of branches per plant with seed yield per plant; number of capsules per plant with 1000 seed weight in cross 1. Number of capsules per plant with number of seeds per capsule also recorded a shift from significant negative to significant positive correlation in the cross 2.

Number of seeds per capsules with seed yield per plant in cross 1 showed negative association in F_3 changed into a significant positive association in the Bips.

Changes in character associations particularly in desirable directions in the biparental progenies were of practical significance in crop improvement. Perhaps an ideal plant could be realized. These associations will be of much use in selecting the desirable traits simultaneously to have a final yield in sesame. The changes might have occurred due to the reshuffling of genes, which could release more additive gene action by forced recombination through intermating (Moll and Robinson, 1967). In such condition, phenotypic selection could uplift stagnant yield levels (Balyan and Verma, 1985).

Joshi (1979) reported that the intermating in selected plants of F_2 resulted in accumulation of favourable genetic effects and break linkages. The breakage would change the magnitude (or) the sign of correlation coefficients between pairs of characters. Therefore, it is evident that in biparental mating result due to breakage of linked gene.

This bring the changes from coupling to repulsion phase and *vice versa*, resulting in reshuffling of genes which could change the association either in positive (or) negative direction (Miller and Rawlings, 1967; Gill *et al.*, 1973; Verma *et al.*, 1979; Yunus and Paroda, 1983).

Since, the Bips generated in the present investigation showed desirable association as well as undesirable associations, selection may be resorted to after another cycle of intermating, which may lead to further improvement by breaking the linkage of undesirable and desirable combination. Hanson (1959) and Krishnadoss and Kadambavanasundaram (1987) had also suggested that the biparental mating followed by recurrent selection was the most effective method for yield improvement in sesame.

Table 126 : Comparison of correlation coefficient among characters between F_3 and Bips

S. No.	Characters	Correlation	
	CO 1 × AUS 107	**F_3**	**Bips**
1.	Plant height *Vs* seed yield per plant	-0.2333*	0.3202**
2.	Number of branches per plant *Vs* seed yield per palnt	-0.3278**	0.0991*
3.	Number of seeds pre capsule *Vs* seed yield per plant	-0.0198	0.6585*
4.	Number of capsules per plant *Vs* 1000 seed weight	-0.9099**	0.2743**
	Paiyur 1 × AUS 107		
1.	Number of capsules per plant *Vs* number of seeds per capsule	-0.8552**	0.2096**

*Significant at 5 per cent; ** Significant at 1 per cent
(Source: Deepalakshmi, 2002)

References

Agarwal, K.B. and R.K. Sharma. 1987. Diallel analysis of duration in rice. Indian J. Genet., 47(1): 84-89.

Al-Jibouri, H.A., P.A. Miller and H.F. Robinson. 1958. Genotypic and environmental variances and co-variance in upland cotton cross of interspecific origin. Agron. J., 50: 633-636.

Allard, R.W. and A.D. Bradshaw. 1964. Implications of genotype ´ environment interactions in applied plant breeding. Crop Sci., 41: 503-508.

Allard, R.W.1960. Principles of plant breeding. Wiley and Sons, Inc., New York.

Altman, D.W. and R.H. Busch. 1984. Random intermating before selection in spring wheat. Crop Sci., 24: 1085-1089.

Ames-Gottifred, N.P. and B.R. Christe. 1989. Competition among strains of *Rhizobium leguminosarum* biovar. *trifoli* and use of a diallel analysis in assessing competition. Appl. Environ. Microbiol., 55: 1599-1604.

Anandan, A., K. Koodalingam and T.S. Raveendran. 2005. Phenotypic stability for yield and its components related characters in hybrids of upland cotton. Agrl. Sci. Dig., 25(1): 59-61.

Anandan, A., R. Eswaran and M. Prakash. 2011. Diversity in rice genotypes under salt affected soil based on multivariate analysis. Pertanika J. Trop. Agrl. Sci., 34(1): 33-40.

Anderson, E. 1939. Recombination in species crosses. Genetics, 14: 668-698.

Andrus, C.F. 1963. Plant breeding systems. Eyphytica, 12: 205-252.

Ansingkar, A.S., V.G. Reddy, D.G. More and M.S. Gaggod. 1992. Feasibility of three-way crosses for exploitation of heterosis in Asiatic cotton. AICCIP Silver Jubilee Symposium, p: A-10.

Arunachalam, V. 1976. Evaluation of diallel crosses by graphical and combining ability methods. Indian J. Genet., 36: 336-358.

Arunachalam, V. and A. Bandyopadhyay. 1984. Limits to genetic divergence for occurrence of heterosis - Experimental evidence from crop plants. Indian J. Genet., 44: 548-554.

Arunachalam, V. and Balarami Reddy. 1979. Basic studies on triallel crosses in pearl millet (*Pennisetum typhoidae* S&H). Z. Pflanzenzijchtg., 83: 368-374.

Arunachalam, V., A. Bandhyopadhyay, S.N. Nigam and R.N. Gibbons. 1984. Heterosis in relation to genetics divergence and specific combining ability in groundnut (*Arachis hypogaea* L.). Euphytica, 33: 33-40.

Arunachalam, V., Bandyopadhyay and M.V. Koteswara Rao. 1985. Performance of three-way crosses in groundnut. Indian J. Agric. Sci., 55: 75-81.

Backiyarani, S. 1995. Genetic analysis of yield and physiological traits in sesame (*Sesamum indicum* L.). Ph.D. Thesis (Unpubl.), AC&RI, TNAU.

Bains, S.S. 1971. Genetics of yield and fibre quality in an intervarietal cross in *Gossypium hirsutum* L. Ph.D. Thesis, I.A.R.I., New Delhi, India.

Baker, F.J. 1969. Genotype ´ environment interactions in yield of wheat. Canadian J. Plant Sci., 49: 743-751.

Baker, R.J. 1978. Issues in diallel analysis. Crop Sci., 18: 533-536.

Balyan, H.S. and A.K. Verma. 1985. Relative efficiency of two intermating systems and selection procedures for yield improvement in wheat (*Triticum aestivum* L.). Theor. Appl. Genet., 71: 111-118.

Bhullar, G.S., K.S. Gill and A.S. Khehra. 1979. Combining ability analysis over F_1 - F_5 generation in diallel crosses of bread wheat. Theor. Appl. Genet., 55: 77-80.

Blackith, R.E. 1960. A synthesis of multivariate techniques to distinguish patterns of growth in grasshoppers. Biometrics, 16: 28-40.

Borges, F.O.L. 1987. Diallel analysis of maize resistance to sorghum downy mildew. Crop Sci., 27: 178-180.

Bray, R.A. 1971. Quantitative evaluation of the circulant partial diallel cross. Heredity, 27: 189-204.

Breese, E.J. 1969. The measurement and significance of genotype ´ environment interaction in grasses. Heredity 24: 27-44.

Bucio-Alanis, L., J.W. Perkins and J.L. Jinks. 1969. Environmental and genotype environmental components of variability V. Segregating generations. Heredity, 24: 115-127.

Bui Chi Buu and Phung Ba Tao. 1992. Genetic nature of some agronomic traits in two elite lines of rice. IRRN, 17: 5.

Buiatti, M., S Baroncelli, A. Bennici, M. Pagliai and R. Tesi. 1974. Genetics of growth and differentiation *in vivo* of *Brassica oleracea* var. *botrytis*. II and *in vitro* and *in vivo* analysis of a diallel cross. Z. Pflanzenzucht, 72: 269-274.

Burton, G.W. 1952. Quantitative inheritance in grasses. Proc. 6th Int. Grassland Cong., 1: 277-283.

Cavalli, L.L. 1952. An analysis of linkage in quantitative inheritance Ed. E.C.R. Rieve and Waddingaton, C.H., HMSO, London. pp. 135- 144.

Chaudhary, B.D. and V.P. Singh. 1976. Triallel analysis for the number of spikes in barley. Crop Improv., 3(1&2): 1-8.

Chaudhary, B.S. and R.S. Paroda. 1980. Phenotype stability for protein content in homogeneous and heterogeneous populations in wheat. Indian J. Genet., 40: 127-131.

Chauhan, V.S., J.S. Chauhan and J.P. Tandon. 1993. Genetic analysis of grain number, grain weight and grain yield in rice. Indian J. Genet., 53(3): 261-263.

Chawla, H.S. and V.P. Gupta. 1983. Seasonal effects on combining ability in pearl millet. Indian J. Genet., 43(2): 137-142.

Clark, A.G. 1987. Senescence and the genetic correlations hang up. Am. Nat., 129: 932-940.

Clegg, M.T., R.W. Allard and A.L. Kahlar. 1972. Is the gene a unit of selection. Evidence from two experimental plant populations. Proc. Nat. Acad. Sci., USA, 69: 2474-2478.

Comstock, R.E. and R.H. Moll. 1963. Genotype ´ environment interactions. Statistical genetics and plant breeding. Nat. Acad. Sci. Nat. Res. Council Publ., 982: 16-196.

Coughtrev, A. and K. Mather. 1970. Interaction and gene association and dispersion in diallel crosses where gene frequencies are unequal. J. Heredity, 35: 69-79.

Cross, H.Z. 1977. Inter-relationships among yield stability and yield components in early maize. Crop Sci., 17: 741-745.

Crumpacker, D.W. and R.W. Allard. 1962. A diallel cross analysis of heading date in wheat. Hilgardia, 32: 275-318.

Curtis, L.C. 1941. Comparative earliness and productiveness of first and second generation summer squash (*Cucurbita pepo*) and the possibilities of using second generation of commercial planting. Proc. American Soc. Hort. Sci., 38: 596-98.

Darlington, D.C. and K. Mather. 1949. The Elements of Genetics. Aller and Unwin, London, p. 446.

Deepalakshmi, C.N. 2002. Studies on genetics of certain economic characters in BIPs populations through North Carolina Design-I in sesame (*Sesamum indicum* L.). M.Sc., (Ag.) Thesis, Annamalai University, Annamalainagar, India.

Delogu, G., C. Corenzoni, A. Marocco, P. Martiniello, M. Oduardi and A.M. Stanca. 1988. A recurrent selection programme for grain yield in winter barley. Euphytica, 37: 105-110.

Dewey, D.R. and K.H. Lu. 1959. A correlation and path coefficient analysis of components of crested wheat grass seed production. Agron. J., 51: 515-518.

Dhanakodi, C.V. and M. Subramanian. 1998. Genetic studies on yield and its component characters in short duration rice varieties. In: First Natl. Plant Breed. Cong., Coimbatore. Abstract, pp. 167-168.

Dhanokodi, C.V. 1990. Genetic studies of yield and its component characters in short duration rice (*Oryza sativa* L.) varieties. Madras Agric. J., 81(6): 313-315.

Dhillon, B.S. 1975. The application of partial-diallel crosses in plant breeding: A review. Crop Improv., 2: 1-7.

Doku, J.L. 1970. Variability in local and exotic varieties of cowpea (*Vigna unguiculata* (L.) Walp.) in Ghana. Ghana J. Agric. Sci., 3: 139-143.

Dwivedi, S.L. and R.B. Singh. 1980. Effects of intermating in early segregating generations on character associations in wheat. Indian J. Agric. Sci., 50: 128-131.

Dwivedi, S.L., K.N. Rai and R.B. Singh. 1980. Diallel analysis of heading date in rice (*Oryza sativa* L.). Theor. Appl. Genet., 57: 43-47

Eberhart, S.A. and C.O. Gardner. 1966. A general model for genetic effects. Biometrics, 22 : 864-881.

Eberhart, S.A. and W.A. Russell. 1966. Stability parameters for comparing varieties. Crop Sci., 6: 36-40.

Eswaran, R. 2007. Heterosis breeding in bhendi (*Abelmoschus esculentus* (L.) Moench.). Ph.D. Thesis, Annamalai University, Annamalainagar, India.

Feyt, H. 1976. Edude critique de analyse des croisements diallel les an moyen de la simulation. Ann. Amelior. Plantes, 26: 173-193.

Finker, V.C., C.G. Porelirt and D.L. Davis. 1973. Heritability of rachis node number of *Avena sativa* L. Crop Sci., 13: 186-188.

Finlay, K.W. 1963. Adaptation – its measurement and significance in barley breeding. Int. Barley Genet. Symp. Proc., 1: 351-359.

Finlay, K.W. 1971. Breeding for yield in barley. Int. Barley Genet. Symp. Proc., 2: 338-345.

Finlay, K.W. and G.N. Wilkinson. 1963. The analysis of genotype ´ environment interactions. Heredity, 31: 339-354.

Fisher, R.A. 1918. The correlation among relatives on the supposition of mendalian inheritance. Trans. Royal Soc. Edinburg, 52: 399-433.

Fisher, R.A. 1936. The use of multiple measurement in taxonomic problems. Ann. Eguenics, 7: 179-178.

Fonseca, A. and F.L. Patterson. 1968. Hybrid vigour in a seven parent diallel cross in common winter wheat (*Triticum aestivum* L.). Crop Sci., 8: 85.

Freeman, G.H. 1973. Statistical methods for analysis of adaptation in plant breeding programmes. Aust. J. Agric. Res., 14: 742-754.

Frey, K.J. 1984. Breeding approaches for increasing cereal crop yields. In cereal production. *In: Cereal production* (Grallagher, E.J. ed.), Buttersworth, London, pp. 47-69.

Gamble, E.C. 1962. Gene effects in corn *Zea mays* L. III. Relative stability of the gene effects in different environments. Canadian J. Pl. Sci., 42: 626-634.

Ganapathy, S. 1989. Studies on the inheritance of yield components in early duration rice (*Oryza sativa* L.). Ph.D. Thesis, TNAU, Coimbatore.

Ganesan, J. 1995. Residual heterosis in sesame (*Sesamum indicum* L.). Plant Breed. Newsletter, 4(2): 3-4.

Gardner, C.O. 1963. Estimates of genetic parameters in cross-fertilizing plants and their implications in plant breeding. *In:* Statistical Genetics and Plant Breeding. Natl. Acad. Sci., Natl. Res. Coun., Washington, pp. 242-252.

Gardner, C.O. and S.A. Eberhart. 1966. Analysis and interpretation of the variety cross diallel and related populations. Biometrics, 22: 439-452.

Gilbert, N.E.G. 1958. Diallel cross in Plant breeding. Heredity, 12: 477-492.

Gill, K.S., G.S. Bhullar, G.S. Mahal and A.S. Khehra. 1977. Combining ability for sedimentation and pelshenke value in F_1 - F_5 diallel crosses in wheat (*Triticum aestivum* L.). Z. Pflanzenzucht, 78: 73-78.

Gill, K.S., S.S. Bains, G. Singh and K.S. Bains. 1973. Partial diallel test crossing for yield and its components in *Triticum aestivum* L. *In:* Proc. 4th Intern. Wheat Genet. Symp., Missouri, Columbia, USA, pp. 29-32.

Giridharan, S. 1986. Diallel, triallel and quadrilateral analysis for cob characters in maize (*Zea mays* L.). Ph.D. Thesis, (Unpubl.) TNAU, Coimbatore.

Gomaa Gibral, A., A. Boe, W.R. Simpson and D.O. Everson. 1982. Evaluation of F_1 hybrid tomato cultivars for earliness, fruit size and yield using diallel analysis. J. Am. Soc. Hort. Sci., 107: 243-247.

Goulden, C.H. 1959. Methods of statistical analysis. Asia Publishing House, New Delhi.

Grafius, J.E. 1956. Components of yield in oats – a geometrical interpretation. Agron. J., 48: 419-423.

Griffing, B. 1956a. A generalized treatment of the use of diallel crosses in quantitative inheritance. Heredity, 10: 31-50.

Griffing, B. 1956b. Concept of general and specific combining ability in relation to diallel crossing systems. Aust. J. Biol. Sci., 9: 463-493.

Guo, P.Z. and S.Z. Wu. 1988. Analysis of gene effects of quantitative characters in rice. Acta Agron. Sin., 14(4): 273-278.

Gurdev Singh, G.S. Bhullar and K.S. Gill. 1986. Comparison of variability generated following biparental mating and selfing in wheat. Crop Improv., 13(1): 24-28.

Hallauer, A.R. 1981. Selection and breeding methods. In Plant Breeding II (Frey, K.J. ed.,), Iowa State Univ. Press, America, pp. 3-55.

Hauller, A.R. 1986. Selection and breeding methods. In: Plant Breeding II (Frey, K.J. ed.) Iowa State University Press, Ames, pp. 3-55.

Hanson, W.D. 1959. The breakup of initial linkage blocks under selected mating systems. Genetics, 44: 857-868.

Hanson, W.D. 1961. Heritability in statistical genetics plant breeding. Nat. Acad. Sci., Nat. Res. Coun., Washington, 125-140.

Hanson, W.D., A.H. Probst and B.E. Chodwell. 1967. Evaluation of population of soybean genotypes with implications from improving the self pollinated crops. Crop Sci., 7: 99-102.

Hayes, H.K. and F.R. Immer. 1942. Methods of Plant Breeding. McGraw-Hill, New York.

Hayman, B.I. 1957. Interaction, heterosis and diallel crosses. Genetics, 42: 336-355.

Hayman, B.I. 1954a. The analysis of variance of diallel tables. Biometrics, 10: 235-244.

Hayman, B.I. 1954b. The theory and analysis of diallel crosses. Genetics, 39: 789-809.

Hayman, B.I. 1958. The theory and analysis of diallel crosses-II. Genetics, 43: 63-85.

Hayman, B.I. and K. Mather. 1955. The description of genetic interaction in continuous variation. Biometrics, 11: 69-82.

Hayward, W.D. 1979. The application of the diallel cross to outbreeding species. Euphytica, 28: 729-737.

Hazel, L.N. and J.L. Lush. 1943. The efficiency of three methods of selection. J. Hered., 33: 393-399.

Hussain Sahib, K. and B.B. Reddy. 1989. Comparison of single, three-way and double crosses in sorghum. J. Res. APAU., 17: 18-23.

Jagtap, D.R. and A.K. Kolhe. 1987. Graphical analysis of yield and its components in upland cotton. ISCI. J., 9: 77-82.

Jensen, N.E. 1970. A diallel selective mating system for cereal breeding. Crop Sci., 10: 629-635.

Jeswani, L.M., B.R. Murthy and R.E. Mehra. 1970. Divergence in relation to geographical origin in a world collection of linseed. Indian J. Genet., 30: 11-25.

Jinks, J.L. 1954. The analysis of continuous variation in a diallel cross of *Nicotiana rustica* varieties. Genetics, 39: 767-788.

Jinks, J.L. and B.I. Hayman. 1953. The analysis of diallel crosses. Maize Genetic Coop. Newsl., 27: 48-54.

Jinks, J.L. and J.M. Perkins. 1970. A general method for the detection of additive, dominance and epistatic components of variation III. F_2 and back cross populations. Heredity, 25: 419-429.

John Joel, A.J. 1987. Multivariate analysis in sesame (*Sesamum indicum* L.). M.Sc. (Ag.) Thesis, submitted in Tamil Nadu Agricultural University, Coimbatore, India.

Johnson, H.W., H.F. Robinson and R.E. Comstock. 1955. Estimates of genetic and environment variability in soybean. Agron. J., 47: 314-318.

Johnson, L.P.V. 1963. Application of the diallel cross technique in plant breeding. State Genet. Plant Breeding, NAS-NRC, 982: 561-570.

Johnson, V.A., J.L. Shafer and J.W. Schmidt. 1968. Regression analysis of general adaptation in hard red winter wheat (*Triticum aestivum* L.). Crop Sci., 8: 186-191.

Joshi, A.B. 1979. Breeding methodology for autogamous crops. Indian J. Genet., 39: 567-578.

Joshi, A.B. and N.L. Dhawan. 1966. Genetic improvement in yield with special reference to self-fertilizing crops. Indian J. Genet., 26: 101-113.

Joshi, A.K. 1990. Triallel analsis in wheat. Crop Improv., 17: 184-185.

Kempthorne, O. 1957. An introduction of genetic statistics. John Wiley and Son., Inc., New York.

Kempthorne, O. and R.N. Curnow. 1961. The partial diallel cross. Biometrics, 17: 229-250.

Khadr, F.H. and K.J. Frey. 1965. Effectiveness of recurrent selection in oats breeding (*Avena sativa* L.). Crop Sci., 5: 349-354.

Koevering, M.V., K.Z. Haufler, D.W. Fulbright, T.G. Isleib and E.H. Everson. 1987. Heritability of resistance in winter wheat spindle streak mosaic virus. Phytopathology, 77: 742-744.

Koodalingam, K. 1994. Genetics of resistance to gall midge and yield component in medium duration rice (*Oryza sativa* L.). Ph.D. Thesis, TNAU, Coimbatore.

Krishnadoss, D. and M. Kadambavanasundaram. 1987. Correlation between yield and yield components in sesame. J. Oilseeds Res., 3: 205-209.

Krueger, S.K., A.A. Weinman and W.H. Gabelman. 1989. Combining ability among inbred onions for resistance to *Fusarium* basal rot. Hort. Sci., 24: 1021-1023.

Langham, D.G. 1946. Genetical studies of the sesame flower. Science, 103: 280.

Larson, R.E. and T.M. Currance. 1937. The extent of hybrid vigour in F_1 and F_2 generation of tomato crosses with particular reference to early and total yield. Teach. Bull. Minn. Agric. Exp. Sta., 164: 1-31.

Lefort-Buson, 1987. Heterosis and genetic distance in rape seed (*Brassica napus* L.): Use of kinship coefficient. Genome, 29: 11-18.

Lush, J.L. 1940. Intra-sire correlation and regression of offsprings on daws as a method of estimating heritability of characters. Proc. Amer. Soc. Animal Production, 33: 293-301.

Lush, J.L. 1945. Animal Breeding Plans, 3rd edition, Iowa State Univ. Press, America.

Mahalanobis, P.C. 1936. On the generalized distance in statistics. Proc. Nat. Inst. Sci., India, 2: 49-55.

Mather, K. 1943. Polygenic inheritance and natural selection. Biol. Revs. Cambridge Phil. Soc., 18: 32-64.

Mather, K. 1949. Biometrical genetics, Methuen and Co. Ltd., London, p. 162.

Mather, K. and J.L. Jinks. 1971. Biometrical genetics. 2nd edition, Chapman and Mill Ltd., New Fetter lane, London.

Mather, K. and J.L. Jinks. 1977. Introduction to biometrical genetics. Chapman and Hall, London.

Mather, K. and J.L. Jinks. 1982. Biometrical genetics. 2nd edn., Chapman and Hall, London.

Matzinger, D.F., G.F. Spraque and C.C. Cockerham. 1959. Diallel crosses of maize in experiments repeated over locations and years. Agron. J., 51: 346-350.

Matzinger, D.F. and E.A. Wernsman. 1968. Four cycles of mass selection in synthetic variety of an autogamous species of *Nicotiana tobacum*. Crop Sci., 8: 239-243.

Matzinger, D.F., G.F. Spraque and C.C. Cockerham. 1969. Diallel crosses of maize in experiments repeated over locations and years. Agron. J., 51: 346-350.

Matzinger, D.F., T.J. Menz and H.F. Robinson. 1960. Genetic variability in flue-cured varieties of *Nicotiana tabacum* L. Agron. J., 52: 8-11.

Mehra, R.B. and S. Ramanujam. 1979. Adaptation in segregation populations of Bengal gram. Indian J. Genet., 39: 492-500.

Miller, P.A. and J.O. Rawlings. 1967. Breakup of initial linkage blocks through intermating in a cotton population. Crop Sci., 7: 199-204.

Moll, R.H. and C.W. Stuber. 1971. Comparison of response to alternative selection procedures initiated with two population of maize (*Zea mays* L.). Crop Sci., 11: 706.

Moll, R.H. and H.F. Robinson. 1967. Quantitative genetic investigation of yield of maize. Zuchter, 37:192-199.

Morishima, H. and H.I. Oka. 1960. The patterns of interspecific variation in the genus *Oryza*: its quantitative representation by statistical methods. Evolution, 14: 153-165.

Morley-Jones, R. 1965. Analysis of variance on the half diallel table. Heredity, 20: 117-121.

Murthy, B.R. 1965. Heterosis and combining ability in relation to genetic divergence in flecured tobacco. Indian J. Genet., 25: 46-56.

Murthy, B.R. 1971. Development traits in breeding for disease resistance in some cereals. *In:* Mutation Breeding for Disease Resistance, IAEA, Vienna, pp. 93-105.

Murthy, B.R. and V. Arunachalam. 1966. The nature of divergence in relation to breeding system in some crop plants. Indian J. Genet., A: 188-198.

Murthy, B.R. and Ziauddin Ahmad. 1974. Components of genetic variation in biparental progenies of a diallel set in dwarf *Penniseums*. Crop Improv., 1(1-2): 82-91.

Murthy, B.R., G.S. Murthy and M.V. Pavats. 1962. Studies in quantitative inheritance in *Nicotiana tobacum*. Components of genetic variation for flowering time, leaf number, grade performance and leaf burn. Zuchter, 361-369.

Muthukumaran, M. 1993. Effect of biparental mating on genetic variability and character association in sesame (*Sesamum indicum* L.). M.Sc. (Ag.) Thesis, Tamil Nadu Agric. Univ., Coimbatore.

Nadarajan, N. and S. Kumaravelu. 1994. Characters association and component analysis in rice under drought stress. Oryza, 31(4): 309-310.

Nanda, G.S., Gurdev Singh and K.S. Gill. 1990. Efficiency of intermating in F_2 generation of an intervarietal cross in breadwheat. Indian J. Genet., 50(4): 364-368.

Nandpuri, K.S., Surjan Singh and Tarsemlal. 1973. Studies on the genetic variability and correlation of economic characters in tomato. J. Res. PAU, 10: 316-321.

Narasimman, R. 2005. Genetics of earliness in rice (*Oryza sativa* L.). Ph.D. Thesis, Annamalai University, Annamalainagar, India.

Nienhuis, J. and S.P. Singh. 1986. Combining ability analysis and relationships among yield, yield components and architectural traits in dry bean. Crop Sci., 26: 21-27.

Nizama, J.R., V.N. Shroff, U.G. Patel and A.R. Dabholkar. 1988. Phenotypic stability of three way hybrids of cotton with respect to their yield and yield components. ISCI J., 13(2): 118-123.

Panse, V.G. and P.V. Sukhatme. 1961. Statistical methods for agricultural workers. ICAR, New Delhi, p. 381.

Pant, D., S.N. Mishra and J.S. Verma. 1992. Effects of biparental mating in forage oats. Indian J. Genet., 53(3): 270-274.

Parameswari, C., V. Muralidharan and B. Subbalakshmi. 1998. Variability studies of the intermated progenies of sesame (*Sesamum indicum* L.). Sesame and Safflower Newsl., 13: 49-54.

Paroda, R.S. and J.D. Hayes. 1971. An investigation of environment interactions for rate for ear emergence in barley. Heredity, 26: 157-175.

Parthasarathy, P. 1994. Diallel and triallel analyses for yield and yield components in rice (*Oryza sativa* L.). Ph. D. Thesis (Unpubl.), TNAU, Coimbatore.

Perkins, J.M. and J.L. Jinks. 1968a. Environmental and genotypic environment components of variability III. Multiple lines and crosses. Heredity, 23: 339-356.

Perkins, J.M. and J.L. Jinks. 1968b. Environmental and genotype ´ environmental components of variability. IV. Non-linear interactions of multiple inbred lines. Am. Fotato J., 36: 381-385.

Plaisted, R.L. and L.C. Peterson. 1959. A technique for evaluating the ability of selection to yield consistently in different locations and seasons. Amer. Potato J., 36: 381-385.

Ponnuswamy, K.N. 1972. Some contribution to design and analysis for diallel and triallel crosses. Ph.D. Dissertation, IARI, New Delhi.

Ponnuswamy, K.N., M.N. Das and M.I. Handoo. 1974. Combining ability analysis for triallel cross in maize (*Zea mays* L.). Theor. Appl. Genet., 45: 170-175.

Pooni, H.S., J.L. Jinks and R.K. Singh. 1984. Methods of analysis and the estimation of the genetic parameters from a diallel set of crosses. Heredity, 52: 243-253.

Pooran Chand and C. Raghunadha Rao. 2001. Impact of different mating approaches in generating variability in blackgram (*Vigna mungo* (L.) Hepper). Legume Res., 24(3): 174-177.

Powell, W. 1988. Diallel analysis of barley anther culture response. Genome, 30: 152-157.

Pundir, R.P.S. and B. Raj. 1971. Multiple selection criteria in the genetic improvement of Toria populations. Indian J. Genet. Pl. Breed., 31(2): 377-382.

Radwan, S.R.H. and A.A. Abo El-Zahab. 1974. Diallel analysis of some agronomic characters in correlation studies in *Sesamum indicum* L. Indian J. Agrl. Res., 15: 119-122.

Ram, T. 1994. Genetics of yield and its components in rice (*Oryza sativa* L.). Indian J. Genet., 54(21): 149-154.

Ram, T., J. Singh and R.M. Singh. 1989. Genetics and order effects of seed weight in rice - a triallel analysis. J. Genet. and Breeding, 44: 53-58.

Ram, T., J. Singh and R.M. Singh. 1990. Estimation of gene effects and order of the parents in three way crosses for certain characters (days to panicle emergence, panicle length and flag leaf area) in rice (*Oryza sativa* L.). Oryza, 27: 385-392.

Ram, T., J.Singh and R.M. Singh. 1994. Analysis of gene effects, combining ability and order of the parents in three way crosses in rice (*Oryza sativa* L.) for number of grains per panicle and grain yield. Oryza, 31: 1-5.

Ramage, R.T. 1981. Comments about the age of male sterile facilitated recurrent selection. Barley Newsletter, 24: 52-53.

Ramanujam, S. and D.K. Tirumalachari. 1967. Genetic variability of certain characters in red pepper. Mysore J. Agric. Sci., 1: 30-36.

Ranga Rao, V., M. Ramachandran and Jawahar Ram Sharma. 1980. Multivariate analysis in genetic divergence in Safflower. Indian J. Genet., 40: 72-89.

Rao, C.R. 1952. Advanced statistical methods in biometric research, John Wiley and Sons, New York.

Rao, C.R. 1958. Bengal anthropometric survey. 1945. A statistical study. Sankhya, 19: 201-408.

Rawlings, J.O. and C.C.Cockerham. 1962. Triallel analysis. Crop Sci., 2: 228-231.

Redden, R.J. and N.P. Jensen. 1974. Mass selection and mating system in cereals. Crop Sci., 407-409.

Riccharia, A.K. and R.S.Singh. 1983. Heterosis in relation to *per se* performance and effects of general combining ability in rice. In: Pre-congress Scientific Meeting on genetics and improvement of heterotic systems, May 5-8 Tamil Nadu Agrl. Univ., Coimbatore, India.

Robin, S. 1997. Genetic analysis of yield, yield components and physiological attributes related with drought tolerance in rice (*Oryza sativa* L.). Ph.D. Thesis, AC & RI, TNAU, Madurai.

Robinson, H.F. 1966. Quantitative gentics in relation to breeding on the centennial of mendelism. Indian J. Genet., 26: 171-187.

Rojas, B.A. and G.F. Sprague. 1952. Comparison of variance components in corn yield trials. III. General and specific combining ability and their interactions with locations and years. Agron. J., 44: 462-466.

Roy, A. and D.V.S. Panwar. 1993. Nature of gene interaction in the inheritance of quantitative characters in rice. Ann. Agric. Res., 14(3): 286-291.

Samuel, C.J., J. Hill, E.L. Brees and A. Davis. 1970. Assessing and predicting environmental response in *Lolium perenne.* J. Agric. Sci. Cam., 81: 1-9.

Saravanan, K. 2001. Studies on genetics of heterosis in bhendi (*Abelmoschus esculentus* (L.) Moench.). Ph.D. Thesis, Annamalai University, Annamalainagar, India.

Senthil Kumar, N. 1998. Genetic studies on certain quantitative inherited characters in two intervarietal crosses of bhendi (*Abelmoschus esculentus* (L.) Monech.). M.Sc. (Ag.) Thesis, Annamalai Univ., Annamalai Nagar.

Shaalan, M.I. and A.E. Aly. 1977. Studies of types of gene effects in some crosses of rice (*Oryza sativa* L.). Libyan J. Agric., 6: 235-243.

Shanthi, R.M. 1989. Studies on biparental mating in rice (*Oryza sativa* L.). M.Sc. (Ag.) Thesis, Tamil Nadu Agricultural University, Coimbatore.

Sharma, J.R., S.P. Yadav and J.N. Singh. 1979. The components of genetic variation in biparental progenies and their use in breeding of pearl millet (*Pennisetum tyhpoides* Burm. S & H). Z. Pflanzenzuchtg, 82: 250-257.

Shroff, V.N., S. Dubey, Rajesh Fulka, S.C. Pandey, S.B. Jadhav and K.C. Mandloi. 1983. Evaluation of commercial hybrids from cytoplasmic male sterility in cotton. ISCI J., 8: 5-14.

Shukla, G.K. 1972. Some statistical aspects of partitioning genotype-environmental components of variability. Heredity, 29: 237-345.

Shull, G.H. 1948. What is heterosis? Genetics, 33: 439-446.

Simmonds, N.W. 1979. Principles of crop improvement. Longman Group Ltd., London, p. 408.

Singh, A.L., J.P. Shahi, J.K. Singh and R.N. Singh. 2001. Genetic control of some traits in maize. Crop Improv., 28(1): 56-61.

Singh, B.B. and B.R. Murthy. 1973. A comparative analysis of biparental mating and selfing in pearl millet (*Pennisetum typhoides*). Theor. Appl. Genet., 43: 18-22.

Singh, D. and M.P. Gupta. 1984. Genetic divergence among thirty-one genotypes of toria (*Brassica campestris* L.) grown in twelve environments. Theor. Appl. Genet., 69: 129-131.

Singh, R.B. and M.P. Gupta. 1968. Multivariate analysis of divergence in upland cotton. Indian J. Genet., 28: 151-157.

Singh, R.B. and S.V. Singh. 1980. Phenotypic stability and adaptability of durum and bread wheats for grain yield. Indian J. Genet., 40: 86-92.

Singh, R.K. and B.D. Chaudhary. 1985. Biometrical methods in quantitative genetic analysis. Revised Ed. Kalyani Publishers, New Delhi.

Singh, R.K. and B.D. Chaudhary. 1977a. The order effects in double cross hybrids. Crop Improv., 4: 213-220.

Singh, R.K. and B.D. Chaudhary. 1977b. Biometrical methods in quantitative genetic analysis. Kalyani Publishers, New Delhi.

Singh, R.K. and B.D. Chaudhary. 1979. Biometrical methods in quantitative genetic analysis. Kalyani Publishers, New Delhi.

Singh, S.V., P.K. Gupta and K.P. Sharma. 1988. Genetic divergence in potato (*Solanum tuberosum* L.) grown in twelve environments. J. Indian Potato Assoc., 15: 45-52.

Singh, V.K., H.G.Singh and Y.S.Chauhan. 1983. Combining ability on sesame. Indian J. Agric. Sci., 53: 305-310.

Sivasubramanian, S. and P. Madhava Menon. 1973. Genotypic and phenotypic variability in rice. Madras Agric. J., 60: 1093-1096.

Smith, H.F. 1936. A discriminant function for plant selection. Ann. Eugenics, 7: 240-250.

Snedecor, G.W. and Cochran. 1961. Statistical methods. The Iowa State University Press, Ames, IOWA, USA, pp. 388.

Sprague, G.F. 1966. Quantitative genetics in plant breeding. Plant Breeding (Ed. Frey, K.J.), Iowa State Univ., Iowa.

Sprague, G.P. and L.A. Tatum. 1942. General vs specific combining ability in single crosses of corn. J. Am. Soc. Agron., 34: 923-932.

Stephens, S.G. 1942. Yield characters and selected oat varieties in relation to cereal breeding technique. J. Agric. Sci., 32: 217-253.

Subbaraman, N. 1984. Genetic analysis of yield and early maturity in rice (*Oryza sativa* L.) by different biometrical techniques. Ph.D. Thesis, TNAU, Coimbatore.

Subbalakshmi, B. 1989. A study on the genetic system governing yield and yield components in sesamum (*Sesamum indicum* L.) through diallel and double cross analysis. Ph.D. Thesis, submitted to Tamil Nadu Agricultural University, Coimbatore, India.

Thirugnanakumar, S. 1992. Seed genetics in relation to yield in sesame (*Sesamum indicum* L.). Ph.D. Thesis, submitted to Tamil Nadu Agricultural University, Coimbatore, India.

Thirugnanakumar, S. 2002. Breeding for saline tolerance in gingelly (*Sesamum indicum* L.). Project Report submitted to UGC, New Delhi, India.

Thirugnanakumar, S. 1991. Seed genetics in relation to yield in sesame (*Sesamum indicum* L.). Unpublished Ph.D. Thesis, submitted to Tamil Nadu Agricultural University, Coimbatore, India.

Timothy, D.H. 1963. Genetic diversity in heterosis and use of exotic stocks in maize at columbia. Proc. Statistical Genet. Pl. Breed. Symp. (Raleigh): NAS-NRC Publ., 982: 581-893.

Umakanth, A.V., E. Satyanarayana and M.V. Nagesh Kumar. 2000. Genetic variability in two maize populations. Ann. Agric. Res., 21(2): 230-233.

Ushakumari, R. 1995. Genetic analysis of yield and yield characters in rice hybrids of indica/japonica wide compatible varieties. Ph.D. Thesis, Tamil Nadu Agrl. Univ., Madurai.

Vaidyanathan, P. 1982. Combining ability and character association studies for attributes related to fodder production in sorghum (*Sorghum bicolor* (L.) Moench.). Ph.D. Thesis, submitted to TNAU, Coimbatore, India.

Vaithilingam, R. 1995. Genetic analysis of earliness and yield components in rice (*Oryza sativa* L.). Ph.D. Thesis, AC & RI, TNAU, Madurai.

Verma, M.M., S. Kochhar and W.R. Kapur. 1979. The assessment of biparental approach in a wheat cross. Z. Pflanzenzuchtg, 82: 174-181.

Vijayakumar, M. 1998. Studies on the genetics of certain economic characters in BIPs population through North Carolina Design-I in cotton (*Gossypium hirsutum* L.). M.Sc. (Ag.) Thesis, Annamalai University, Annamalainagar, India.

Wallace, 1963. Models of reproduction and their genetic consequences in statistical genetics and plant breeding. Symp. Raleigh., pp. 3-20.

Walsh, E.J. and R.E. Atkings. 1973. Performance and within hybrid variability of three way and single crosses of grain sorghum. Crop Sci., 13: 267-271.

Walton, P.D. 1976. A genetic study of the factors which constitute forage yield in *Bromus inermis*. Z. Pflanzenzuchtg., 77: 43-45.

Wassimi, N.N., T.G. Isleib and G.L. Hosfield. 1986. Fixed effect genetic analysis of a diallel cross in dry beans (*Phaseolus vulgaris* L.). Theor. Appl. Genet., 72: 449-454.

Weatherspoon, J.H.1970. Comparative yields of single, three way and double cross hybrids in maize. Crop Sci., 10: 157-159.

Williams, W. 1964. Genetical principles and plant breeding. Blackwell Scientific Publishers, Oxford, pp. 1-504.

Williams, W.P. and G.L. Windham. 1988. Resistance of corn to southern root knot nematode. Crop Sci., 28: 495-496.

Wricke, G. and W.E. Weber. 1986. Quantitative genetics and selection in plant breeding de Gruyter, Berlin.

Wright, A.J. 1985. Diallel designs, analysis and reference populations. Heredity, 54: 307-311.

Wright, J.A., A.R.Hullaurer, L.H.Penny and S.A.Eberhart. 1971. Estimating genetic variances in maize by use of single and three-way crosses among unselected inbred lines. Crop Sci., 11 : 690-695.

Wright, S. 1921. Correlation and causation. J. Agric. Res., 20: 557-585.

Wynne, J.C., D.A. Emery and P.W. Rice. 1970. Combining ability estimates in *Arachis hypogaea* L. II. Field performance of F_1 hybrids. Crop Sci., 10: 713-715.

Yadav, S.P. and B.R. Murthy. 1979. Association characters and utility of biparental progenies in bread wheat. Z. Pflanzenzuchtg, 82: 218-229.

Yunus, M. and R.S. Paroda. 1983. Extent of genetic variability created through biparental mating in wheat. Indian J. Genet., 43: 76-81.

Ziauddin Ahamed, R.P., Katiyar and Radhey Shyam. 1980. Genetic divergence in triticale. Indian J. Genet., 40: 35-39.

Index

A

Accumulation 57, 215, 217, 221
Adaptability 41
Adaptation 41, 42, 56, 72
Additive 9, 78, 84, 91, 93, 94, 96, 101, 104, 106, 107, 108, 110, 112, 113, 114, 125, 127, 128, 130, 134, 139, 140, 145, 148, 149, 160, 163, 164, 166, 167, 168, 169, 171, 173, 174, 175, 179, 180, 181, 182, 183, 184, 199, 200, 201, 203, 204, 205, 206, 207, 209, 212, 215, 216, 217, 220, 221
Additivity 113, 154, 155, 161, 184, 185
Algebraic 96, 101, 148, 160
Alleles 94, 95, 96, 97, 145, 146, 147, 148, 160, 161
Amphi-directional 156
Anomalies 155
Asymmetrical 155
Autogamous 110

B

Biometrical 78, 179, 199
Biparental 211, 212, 214, 215, 216, 218, 219, 220, 221, 222
Buffering 114, 115

C

Coefficients 11, 12, 13, 14, 16, 18, 20, 21, 23, 24, 25, 26, 27, 28, 29, 30, 31, 42, 45, 54, 55, 100, 153, 154, 220, 221
Combiner 93, 100, 106, 107, 108, 111, 112, 113, 114, 126
Combining 77, 78, 80, 83, 84, 85, 86, 87, 91, 92, 93, 94, 95, 98, 99, 103, 104, 105, 106, 107, 108, 111, 112, 113, 114, 117, 126, 164
Competition 93
Complementary 91, 215
Complete 92, 97, 148, 154, 155, 160, 161, 217

Correlation 11, 12, 13, 14, 16, 18, 20, 21, 31, 45, 53, 54, 55, 153, 154, 156, 212, 220, 221

Covariances 92, 95, 96

Cumulative 115

D

Depression 165, 166, 172

Diallel 78, 91, 92, 93, 95, 97, 103, 112, 113, 114, 121, 129, 133, 134, 140, 141, 147, 148, 151, 160, 177

Digenic 113, 167, 181, 183

Discriminant 33, 34, 35, 36, 37, 38, 39

Divergence 57, 58, 59, 64, 73, 74

Diversity 57, 58, 60, 62, 63, 65, 66, 67, 68, 73, 74, 75, 154, 156, 160, 173

Dominant 48, 91, 96, 97, 102, 112, 113, 147, 154, 155, 156, 160, 161, 176, 177, 183, 184, 185, 186, 217, 218, 220

Donor 35, 37

Duplicate 176, 177, 183, 184, 185, 186

E

Eco-geographic 60, 62

Efficiency 34, 35, 37, 169, 170, 174, 177

Epistasis 9, 94, 95, 112, 114, 125, 146, 147, 148, 154, 164, 169, 176, 177, 181, 183, 184, 185, 186, 199, 200, 201, 202, 204, 205, 206, 207, 217

Evolutionary 58, 73

F

Function 33, 34, 35, 36, 37, 38, 39

G

Genotypic 1, 2, 3, 6, 8, 12, 13, 14, 16, 18, 20, 21, 34, 47, 48, 129, 164, 218, 219

Geographical 63, 73

Germplasm 57

H

Heritability 2, 8, 9, 10, 149, 161, 163, 171, 172, 219, 220

Heterobeltiosis 83, 134, 138, 139, 140, 141, 142

Heterogeneity 42

Heterosis 57, 63, 83, 111, 115, 138, 139, 140, 141, 142, 147, 156, 160, 165, 172, 173

Heterozygosity 160, 215

Heterozygous 57, 96, 101, 148, 161

Homeostasis 56

Homozygotes 95, 122, 164

Hybridisation 169, 179

Hybridization 58, 77, 91, 110, 113, 128, 134

Hybrids 77, 78, 80, 81, 83, 84, 88, 89, 90, 91, 92, 95, 100, 103, 104, 106, 111, 112, 113, 114, 115, 116, 121, 122, 126, 129, 133, 134, 138, 139, 140, 141, 142, 145, 151, 156, 164, 169

I

Incomplete 97, 148, 154, 155, 161

Intercept 152, 154

Inter-cluster 59, 64, 72, 73

Intercorrelation 13, 53

Intersection 96, 97, 103

Intra-cluster 59, 65, 66, 67, 68, 73

L

Linkages 211, 212, 217, 218, 219, 221

M

Mahalanobis 58, 59

Maternal 94

Multiplier 101, 102
Multivariate 58, 59, 72

N

Non-additive 78, 84, 94, 104, 106, 112, 125, 163, 183, 212
Non-allelic 146, 155
Non-linear 48, 53

P

Partial 92, 93, 154, 174, 217
Partitioning 169, 170, 174, 177, 182, 201, 205
Path-coefficient 12
Pedigree 211
Phenotypic 2, 6, 7, 12, 13, 14, 16, 18, 21, 43, 44, 45, 48, 50, 73, 91, 110, 111, 112, 116, 122, 149, 163, 218, 219, 220, 221
Polycross 78
Polygenic 11, 64, 72, 92, 212
Pooled 3, 4, 5, 6, 7, 8, 9, 29, 30, 37, 39, 40, 44, 45, 47, 48, 49, 50, 58, 60, 62, 63, 64, 72, 74, 75, 86, 87, 89, 90
Potentiality 91, 103, 128, 171, 219
Predominant 48, 112, 113, 183, 217, 218, 220
Progeny 91, 92, 94, 101, 110, 149, 163, 213

Q

Quadriallel 112, 128, 131
Quantitative 57, 58, 77, 92, 163, 179, 211, 216, 220

R

Recessive 91, 96, 97, 102, 148, 154, 155, 156, 160, 176, 177, 184, 185, 186
Reciprocal 92, 93, 94, 95, 98, 99, 103, 104, 107, 121, 140, 160, 170, 175, 183, 217
Recombinants 177, 211, 212, 217, 218
Recombination 108, 111, 177, 200, 211, 212, 217, 220, 221
Recurrent 94, 96, 177, 211, 217, 218, 222
Regression 42, 43, 45, 48, 51, 96, 97, 101, 102, 103, 146, 148, 151, 152, 153, 154, 155
Repulsion 200, 212, 217, 218, 219, 220, 222
Reshuffling 221, 222
Residual 23, 24, 25, 26, 27, 28, 29, 30, 165, 172, 173
Resorting 9, 104, 128, 145, 177

S

Segregants 169, 170, 173, 177, 214, 217, 218
Segregation 95, 108, 111, 170
Selection 1, 7, 9, 11, 12, 13, 20, 33, 34, 35, 36, 37, 38, 39, 41, 42, 46, 53, 57, 63, 64, 73, 145, 149, 163, 164, 169, 171, 174, 177, 179, 211, 212, 216, 217, 218, 219, 220, 221, 222
Stability 41, 42, 43, 44, 45, 48, 50, 51, 52, 53, 56, 115
Stimulation 147
Stratification 41
Superiority 170, 214

T

Tolerance 35, 37
Transgressive 170, 173
Transmission 8, 126
Triallel 111, 112, 113, 114, 115, 116, 121, 126, 127, 128
Triple 113, 127, 128, 199, 201, 202

U

Unfavourable 41, 108, 220

V

Variability 1, 2, 8, 42, 103, 108, 110, 121, 128, 149, 163, 164, 171, 172, 211, 214, 217, 218, 219, 220

Variance 1, 2, 3, 4, 6, 7, 8, 44, 45, 47, 49, 50, 80, 81, 82, 83, 84, 85, 86, 88, 94, 95, 96, 98, 99, 100, 101, 102, 103, 104, 105, 110, 112, 114, 118, 121, 125, 128, 132, 133, 134, 138, 140, 145, 148, 149, 163, 164, 165, 166, 168, 169, 180, 182, 183, 185, 199, 200, 201, 203, 204, 205, 206, 207, 212, 213, 216, 217, 218, 219, 220